Internet of Things Security:
Fundamentals, Techniques and Applications

RIVER PUBLISHERS SERIES IN SECURITY AND DIGITAL FORENSICS

Series Editors:

WILLIAM J. BUCHANAN
Edinburgh Napier University, UK

ANAND R. PRASAD
NEC, Japan

Indexing: All books published in this series are submitted to the Web of Science Book Citation Index (BkCI), to CrossRef and to Google Scholar.

The "River Publishers Series in Security and Digital Forensics" is a series of comprehensive academic and professional books which focus on the theory and applications of Cyber Security, including Data Security, Mobile and Network Security, Cryptography and Digital Forensics. Topics in Prevention and Threat Management are also included in the scope of the book series, as are general business Standards in this domain.

Books published in the series include research monographs, edited volumes, handbooks and textbooks. The books provide professionals, researchers, educators, and advanced students in the field with an invaluable insight into the latest research and developments.

Topics covered in the series include, but are by no means restricted to the following:

- Cyber Security
- Digital Forensics
- Cryptography
- Blockchain
- IoT Security
- Network Security
- Mobile Security
- Data and App Security
- Threat Management
- Standardization
- Privacy
- Software Security
- Hardware Security

For a list of other books in this series, visit www.riverpublishers.com

Internet of Things Security:
Fundamentals, Techniques and Applications

Editors

Shishir Kumar Shandilya

VIT Bhopal University
India

Soon Ae Chun

City University of New York
USA

Smita Shandilya

Sagar Institute of Research, Technology and Science
India

Edgar Weippl

SBA Research
Austria

River Publishers

Published, sold and distributed by:
River Publishers
Alsbjergvej 10
9260 Gistrup
Denmark

River Publishers
Lange Geer 44
2611 PW Delft
The Netherlands

Tel.: +45369953197
www.riverpublishers.com

ISBN: 978-87-93609-53-2 (Hardback)
    978-87-93609-52-5 (Ebook)

Contents

Shishir Kumar Shandilya, Soon Ae Chun, Smita Shandilya
and Edgar Weippl

2 Internet of Things Privacy, Security, and Governance

Gianmarco Baldini, Trevor Peirce, Marcus Handte,
Domenico Rotondi, Sergio Gusmeroli, Salvatore Piccione,
Bertrand Copigneaux, Franck Le Gall, Foued Melakessou,
Philippe Smadja, Alexandru Serbanati, and Julinda Stefa

Foreword

The world has witnessed tremendous growth in machine-to-machine (M2M) communication technology in the last decade. As a result of this, a large range of M2M communication techniques have emerged and changed the entire scenario. These M2M-enabled nodes have created a highly sophisticated and hi-tech Internet of thing (IoT) environment for many purposes.

IoT security is collection of tools and techniques for safeguarding connected devices and networks in IoT domain. But, this important thing generally remains unconsidered in the IoT design. The IoT products are often implemented with old or missing security functionalities. Also, the clients are often failed to change even the default passwords of smart devices. Security experts keep warning the organizations and people for the potential risk of having a large number of unsecured devices and breach of privacy due to this. One such security measure to avoid such risks is Identity of Things (IDoT), which assigns unique identifiers (UIDs) to the smart devices as per the associated metadata and then only allows them to communicate with each other. Identity of things is an essential component of the IoT, in which almost anything imaginable can be addressed and networked for exchange of data online.

There are not many books on IoT Security; the few that have published deal mostly only with the installation of secured IoT. This book not only discusses the IoT Security techniques in detail but also tries to provide a comprehensive picture on the subject which is necessary to build highly secure and feature-rich IoT applications.

Finally, I would like to say that the presented book is a comprehensive and well-written guide on the topic. I believe that students and practitioners will especially enjoy reading this work.

Prof. Atulya Nagar holds the Foundation Chair as Professor of Mathematical Sciences and is the Dean of Faculty of Science at Liverpool Hope University in United Kingdom. Professor Nagar is an internationally recognized scholar working at the cutting edge of theoretical computer science, applied mathematical analysis, operations research, and systems engineering, and his work is underpinned by strong complexity-theoretical foundations.

Preface

IoT security is the area of endeavor concerned with safeguarding connected devices and networks in the Internet of things (IoTs). It involves the increasing prevalence of objects provided with unique identifiers and the ability to automatically transfer data over a network. Much of the increase in IoT communication comes from computing devices and embedded sensor systems used in industrial machine-to-machine (M2M) communications, smart energy grids, home and building automation, vehicle-to-vehicle communication, and wearable computing devices.

The main problem is that because the idea of networking appliances and other objects is relatively new, security has not always been considered in the product design. IoT products are often sold with old embedded operating systems and software. To improve security, an IoT device that needs to be directly accessible over the Internet should be segmented into its own network and have network access restricted. The network segment should then be monitored to identify potential anomalous traffic, and actions should be taken if there is a problem. This issue is an open research problem, and only few references are available on this subject.

This book is focused on current research while highlighting the empirical results along with theoretical concept to provide a good comprehensive reference for students, researchers, scholars, professionals, and practitioners in the field of advanced Security and IoT.

We express our heartfelt gratitude to all the authors, reviewers, and River Publishers personnel, especially Mr. Mark De Jongh for their kind support. Special thanks to Professor Atulya Nagar, Dr. Suresh Jain, and

Dr. Ganga Agnihotri, for their endless motivation and patience. We hope that this book will be beneficial to all concerned readers.

Shishir Kumar Shandilya
VIT Bhopal University, India

Soon Ae Chun
City University of New York, USA

Smita Shandilya
Sagar Institute of Research, Technology and Science, India

Edgar Weippl
SBA Research, Austria

List of Contributors

Alexandru Serbanati, *Sapienza University of Rome, Italy*

Bertrand Copigneaux, *Inno TSD, France*

Domenico Rotondi, *TXT e-solutions S.p.A., Italy*

Edgar Weippl, *SBA Research, Austria*

Foued Melakessou, *University of Luxemburg, Luxemburg*

Franck Le Gall, *Inno TSD, France*

Gianmarco Baldini, *Joint Research Centre – European Commission, Italy*

Imran Ali Khan, *Department of Computer Science Engineering, Bansal Institute of Research, Technology and Science, Madhya Pradesh, India*

Julinda Stefa, *Sapienza University of Rome, Italy*

Marcus Handte, *Universität Duisburg-Essen, Germany*

Philippe Smadja, *Gemalto, France*

Pijush Kanti Dutta Pramanik, *Dept. of Computer Science & Engineering, National Institute of Technology, Durgapur, India*

Prasenjit Choudhury, *Dept. of Computer Science & Engineering, National Institute of Technology, Durgapur, India*

Priti Maheshwary, *Department of CSE, Rabindranath Tagore University (formerly known as AISECT University), Bhopal, MP, India*

Salvatore Piccione, *TXT e-solutions S.p.A., Italy*

Sergio Gusmeroli, *TXT e-solutions S.p.A., Italy*

Shaligram Prajapat, *International Institute of Professional Studies, Madhya Pradesh, India; Devi Ahilya University, Indore, Madhya Pradesh, India*

Shishir Kumar Shandilya, *VIT Bhopal University, India*

Smita Shandilya, *Sagar Institute of Research Technology and Science, India*

Soon Ae Chun, *City University of New York, USA*

Tasneem Jahan, *Department of Computer Science Engineering, Bansal Institute of Research, Technology and Science, Madhya Pradesh, India*

Timothy Malche, *Department of CSE, Rabindranath Tagore University (formerly known as AISECT University), Bhopal, MP, India*

Trevor Peirce, *AVANTA Global SPRL, Belgium*

List of Figures

List of Tables

List of Abbreviations

ACL	Access Control Lists
ARM	Association Rule Mining
AVK	Automatic Variable Key
COSO	Committee of Sponsoring Organizations of the Treadway Commission
CPU	Central Processing Unit
DDoS	Distributed Denial-of-Service
DoS	Denial-of-Service
DRAM	Dynamic RAM
DTLS	Datagram Transport Layer Security
ELP	Electronic License Plates
ESN	Electronic Serial Number
ETSI	European Telecommunications Standards Institute
HMAC	Hash Message Authentication Code Technique
IDS	Intrusion Detection Systems
IIoT	Industrial Internet of Things
IMC	In-memory Computing
IMEI	International Mobile Equipment Identity
IoT	Internet of Things
IPS	Intrusion Prevention System
IT	Information Technology
LTE	Long Term Evolution
M2M	Machine-to-Machine Communication
MEC	Mobile Edge Computing
MGT	Mobility Grade Threshold
NIST	National Institute of Standards and Technology
P2P	Peer-to-Peer
PKC	Public Key Cryptography
PKI	Public Key Infrastructure
QoE	Quality of Experience
RAID	Redundant Array of Independent Disks

RAM	Random Access Memory
RAN	Radio and Network
RNC	Radio Network Controller
SKC	Symmetric Key Cryptography
TCO	Total Cost of Ownership
VANET	Vehicular Sensor Networks
VMS	Video Management Software
WDR	Wide Dynamic Range
WWW	World Wide Web

1

IoT Security: An Introduction

**Shishir Kumar Shandilya[1], Soon Ae Chun[2],
Smita Shandilya[3] and Edgar Weippl[4]**

[1]VIT Bhopal University, India
[2]City University of New York, USA
[3]Sagar Institute of Research Technology and Science, India
[4]SBA Research, Austria

With the ability to get connected, the IoT has spread its arena while facilitating the users with more comfort to get connected to several devices at one go, to share data and control. This ability to connect, communicate, and management has opened a wide door of opportunities for futuristic technologies to work collaboratively. These communication links are now coming out from closed secured networks to open public Internet networks. This is making a big security issue for the IoT system as devices are becoming intelligent day by day and their interconnection among them is raising the inevitable possibilities of intrusion and interferences. A bunch of customizable strong IoT protection mechanisms are therefore needed to avoid such compromise on privacy and to safeguard the IoT users' data. The presented chapter analyzes these security issues and discusses the security challenges posed by IoT devices with the approaches for respective solutions.

1.1 Introduction

Organizations are essentially required to be completely convinced regarding the security issues before implementing IoT in the existing system or creating an entirely new system. Therefore, the IoT solution providers face many challenges to create faith on the technology. Every organization visualizes and conceptualizes the IoT deployment differently which creates more restlessness and disbelieve on the appropriateness of security measures.

Most of the vendors are more focused on the solutions they can provide to the organization by the pool of sensors, data collection and analysis servers, and optimization subroutines. They are little less concern about the security issues after implementing the system, which is more important issue. Mere providing a customized suite of compatible electronic components with software services in IoT implementation is certainly not enough for the organization looking out for technology upgrade. Every IoT vendor is aware that security has become the prime most concern of organizations since last few years and they have to provide the IoT solution equipped with safe and reliable operations through a number of firewalls and security protocols. However, there is no common security phenomenon by which they can convince their clients on security issues, rather it would require a more personalized approach with customized security constraints. To make the IoT more effective, the organization should rely on it with confidence which is only possible when the vendors have designed the IoT system and implement the security measures in line with the organization. So, it is also about the psychological faith on the technology and vendors are essentially required to achieve that.

1.2 Security: A Major Concern

Security has been always a chief concern ever since the beginning of computing. People reply more on the technologies which offer them more secured environment to work while protecting their privacy and identity. Since IoT came into picture, the computing scenario has been shifted from stand-alone computing to more flexible collaborative computing. This has raised the security concern once again with intensified treat toward the intrusion. It ranges from individual personal information hacks to interfered financial transactions and spoofing. Intelligent devices which are driven by sophisticated programs are more prone to get ill-programmed. Also, the hand-shaking and common collaborative platforms among these devices increase the probability of compromising the security measures. However, this is not only the opportunity for malicious users but also for programmers to achieve the highest level of security to mitigate these security treats.

Since its evolution, the smart devices of the IoT framework have been efficiently delivering their operations. But they also face certain threats to users and their personal data because of the ever expanding network.

1.2.1 Confidentiality

The IoT framework encompasses interconnection of devices, sensors, information, and software services. Confidentiality refers to the property of ensuring that data or network transactions are readable only by the destination they are meant for. The prime goal of confidentiality is to keep focus on identification of devices, communication, and sensing, and on the services concerning semantics. The identification process handles the task of matching network services to the demand of users. Communication deals with linking objects of heterogeneous nature to the specific set of services. In the sensing process, the information obtained from various smart devices is computed according to the user's demand and is sent to the IoT database or cloud as sensed data. This aggregated block of communication and computation is the processing unit of IoT. IoT networks widely use the Datagram Transport Layer Security (DTLS) protocol for achieving confidentiality. It provides two-way authentication, and its underlying principle could be symmetric encryption and elliptic curve cryptography. Data confidentiality is achieved through the implementation of HTTPS protocol by enabling an encrypted and secure communication path between IoT devices to gateway and from gateway to cloud.

1.2.2 Authentication

Authentication ensures the validity of a user in the IoT network. An authenticated user is identified provided that it possesses the authority of communication among its peers. The session keys are generated using session key distribution schemes to enforce authenticity and access control. Public key infrastructure (PKI) has always been the nerve of Internet security. It ensures the mutual trust and device authentication.

1.2.3 Data Integrity

This property of a secured IoT network deals with the data contained in the devices as well as the data flowing between communicating nodes (Figure 1.1). If the data integrity is compromised, it will consecutively result in the exploitation of network devices and the entire IoT platform. The data in transit require to be protected against modifications. Data integrity can be achieved by using keyed-Hash message authentication code technique (HMAC) whose principle is to keep a shared private key; since it needs a shared private key, it must be protected just like any other cryptographic key.

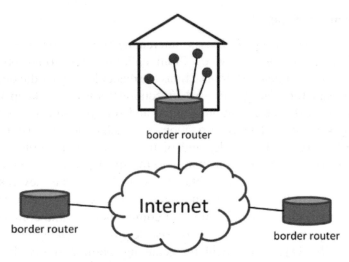

Figure 1.1   IoT network infrastructure.

1.2.4 Cyber Threats and Their Detection

The IoT technology has brought a number of security challenges along with its attractive offerings. From a user's perspective, it has made our daily life activities easier and accurate by smart devices. But from a network's perspective, the expanding nature of IoT networks is vulnerable to powerful cyberattacks.

The cyberattacks can be aimed for:

- Deteriorating physical devices and appliances connected to the IoT network.
- Misusing the incoming and outgoing network traffic.
- Halting the network operation.

Various attack detection techniques can be deployed at routers and switch levels. For example,

- Intrusion prevention systems (IPSs)
- Firewalls
- Intrusion detection systems (IDSs)
- Access control lists (ACLs)

IDS is a potential solution for threat detection and prohibits the misuse of IoT smart devices. The IDS is implemented on the border routers. It monitors the in–out network traffic and alarms upon the detection of malicious activities.

This proactive detection mechanism is, however, not a suitable solution for deep packet inspection.

1.2.5 Threat Mitigation

It is a reactive measure after a threat has been reported. It investigates the impact of a threat on the network and the infected areas of the IoT network. The current IoT networks require models for combined cyber security and physical security. For example, the banking industry has information security that uses Honeypots to know the loopholes that could be targeted points for an attacker. Honeypot focuses on a number of services, such as HTTP, SMPT, SSH, FTP, etc. It offers the advantage of presenting a transparent outlook of prediction of current and future attacks.

1.2.6 Malware Resistance

A malware in an IoT network poses itself as a legitimate network user/device and attempts to authenticate with common username and password. It successfully bypasses the login mechanism, and then raises destructive commands to exploit data integrity and to lead to disruptive device chapters. It could further kill the Internet connection and make the device unusable. It could either over-write device configurations or erase it and may also wipe out the external hard drives.

A Mirai malware launched a destructive attack on October 2016, which was a category of DDoS. It was written in C, and targeted the embedded devices with Linux-based platforms such as-CCTVs, DVRs, routers, etc. It had the property of self-propagation by brute-force telnet passwords. The malware launches malicious code. This code can arrive in a device via spam mail or images and automatically triggers the installation when it opens.

Defense against Mirai:

- By securing IoT devices with a strong password. Timely backups should be done, and network traffic should be captured and analyzed with expert professionals.
- To remotely access Linux accounts, Telnet login should be disabled, and SSH should be used.
- The networks devices are less prone to attacks if are being continuously updated and login credentials are changed with time.
- It is important that strong encryption standards are implemented for the IoT system, so as to prevent their accessibility by attackers.

1.3 Business Aspects in IoT

With the emergence of IoT, the business processes can become smooth and robust at the fundamental level and can reach greater heights with smart devices and automated products. IoT has the power to transform the business applications. The business aspects of IoT have focuses on the following key elements:

- Smarter products in companies: The pocket size smart phones bear a lot of things of the world inside it. The smart devices result in the quick and instantaneous processing of applications and execute business activities in a comparatively short period of time.
- Enable smarter business operations and smarter decisions: The feedback process becomes easy and prediction of future impacts can be judged.
- Change in business model: The primitive producer–consumer trend has been greatly revolutionized by IoT. It offers time-saving business solutions.

1.4 Industrial IoT (IIoT)

The extension of IoT to the manufacturing industry is defined by the term IIoT. It radically changes the process of manufacturing by empowering the addition and accessibility of huge amount of data at extremely high speeds and in a systematic and cost-effective manner. Many of the companies are now implementing IIoT to incorporate sensor data, to constantly and accurately capture the communicating data, and for the efficient monitoring of the overall supply chain, quality control, to store information about equipments, vehicles, and containers and for the traceability of industrial processes.

IIoT produces data which are exponentially large compared to that of generated by IoT. For example, around 500GB data can be produced by a single turbine compressor blade in one single day. And therefore, IIoT requires combining not only computers and Internet but also the modern day technologies of Big Data, Cloud Computing, and Machine Learning. There is also an important remark associated with IIoT, which can be understood as IoT is important but not critical while IIoT failure might result in life-threatening or other emergency situations. IIoT is thus a very promising technology and it assures the emergence of the next industrial revolution by creating a platform of connectivity and to drive industrial solutions with IoT.

The key elements in IIoT are:

- Exchange of information is between business to business, rather than from business to consumers.
- The market segments of IIoT are huge enterprises compared to small business or limited enterprise of IoT.
- The data volume of IoT is termed Big Data. In IIoT, there are limited but specific data.
- IoT focuses on consumer convenience and consumer needs. IIoT aims to cut down the investments and improvement of return costs.

1.4.1 Security Requirements in IIoT

The industrial Internet aims at improving the efficiency and productivity of the production process throughout the supply chain (Figure 1.2). IIoT encompasses those areas with high stake industries, for e.g., oil and gas supply chains, power grids, heavy machineries, and sensors. Any security breach in these applications could cause a huge impact on business solutions. There could be a threat to data security.

Ignoring security and privacy issues could endanger not only the user activities but also the operation and functionality of devices. IIoT security includes the concern for safety and reliability.

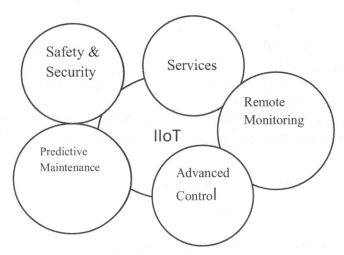

Figure 1.2 IIoT operations.

1.5 Vehicular Sensor Networks (VANETs)

A vehicular *ad hoc* network (VANET) considers a vehicle as a node that transmits messages among vehicles. It is a subset of IoT that could be termed IoV. Vehicles are smart objects, equipped with sensors, and have IP-based connectivity. A VANET transmits messages between intra-vehicle components, vehicle to vehicle, and vehicle to people. The processing of messages is based on sensing capabilities of vehicles.

A VANET boosts the foundation of an Intelligent Transportation System (ITS), which has the potential to offer a rich set of applications to its customers, which acts as a roadside infrastructure for using security and services.

1.5.1 Sensors in VANET

Sensors in VANET can be classified into two categories:

1. Autonomous Sensors: These sensors include acceptance range threshold (ART) and mobility grade threshold (MGT). These parameters take into consideration the maximum communication range and restrict the number of entities in a particular area.
2. Co-operative Sensors: These sensors include steps and measures for practice exchange of neighbor tables among peer vehicles about their data and position.

1.5.2 Security in VANET

The security of a VANET depends on cost, trust, privacy, and its deployment. Certain countries assign electronic license plates (ELPs) to vehicles, which is a cryptographically verifiable number.

The PKI in context of VANET is termed VPKI, which is a certification key distribution, certificate revocation, data recording, etc.

1.6 IoT-Enabled Wearable Devices

The market of wearable smart devices is a widely talked topic in the world of Internet. These devices have a full range of capabilities for various IoT solutions. Current trends of wearable devices have large scope in the health care and fitness sector. Biometric-oriented wearable devices can measure heart rate and oxygen levels in bloodstream, track body temperature, and report the possibility of catching flu or cold. A wearable device could tell if the back seat of a car is not properly adjusted and causing strain on back. In the form of a wristband, it can measure perspiration levels and thus can offer an alarm for adjusting temperature and humidity in an AC room or car. IoT wearable used nowadays are: fitness trackers, smart watches, smart glasses, baby monitoring gadgets, and clothing-based wearables. A survey revealed that 43% will replace their smart phones by wearables.

1.7 IoT in Smart Homes/Cities

A smart city based on the paradigm of IoT uses a number of communication protocols for different data representations. It also utilizes the concepts of artificial intelligence to model the data processing. The smart city or smart home collects data from numerous data sources which are basically sensors. These sensors might be scattered around cities, offices, gardens, public places, and markets. Data arrive from smart phones, smart cards, wearable sensors, vehicles, etc. A smart city IoT network aims for efficient management of energy, water, and electrical supplies in buildings. It also lays down the activities for improving public transport, traffic analysis, and population density statistics. Waste management is also a rising concern in rural and urban areas. This could be successfully done with the aid of smart homes and smart cities.

1.8 Green IoT

Green IoT is a term used for integrating the Greenhouse industry with IoT. It encompasses the procedures which are energy efficient and are utilized to reduce the greenhouse effect of existing systems.

It offers the advantage of managing the following activities:

- Crop management by evaluating farming and environment conditions.
- To improve the yield of crops by prolonged production period and monitoring less use of chemicals and fertilizers.
- To judge and analyze the requirements of soil, water supply, and humidity.
- To monitor the temperature changes.

After the evaluation of climatic and geographic conditions, the green IoT aims to launch the set of activities that would yield better crop production and would enhance the sustainability of farming lands. Not only this, it also launches activities to reduce deforestation and to increase the greener land areas. Certain models of Green IoT use the ZigBee sensor network, for tracking temperature, humidity, and soil control levels. Their threshold levels are stored in the cloud. When the sensor parameters rise above or fall below the threshold, the network raises an alarm. The humidity sensor of a greenhouse generates an alarm upon any change in the humidity level.

1.9 Video Streaming and Data Security from Cameras

The pace with which video data get generated is much faster than that of other data. The surveillance and security cameras constantly generate video data. The video data hold a high asset value for business.

The complex element of management of video data is its unstructured form. The structured video data are easily manageable. Companies and business processes focus on using Video Management Software (VMS) to search these big data for analytical processing of statistics. This processing is based on times, locations, people, and certain keywords. A constraint should be kept in mind regarding loss in the prevention of critical information, during the processing of marketing, operations, and customer service.

Cameras are a useful tool for majority of business applications and in most of the use cases. The wide dynamic range (WDR) property of cameras provides greater details to analytics for deciphering information. With a HD and HDTV camera, the resolution increases to a better range. However, higher resolutions increase storage consumption, and thus require video compression algorithms. The security levels or range of a network camera is analyzed and optimized in real time. The cameras are used to continuously store the information. Certain data in this information might be unimportant and not so useful. The data are filtered according to the purpose. The filtration is done by analytics. Analytics technology is the brain of interconnected IoT devices.

The role of analytics is to evaluate security aspects of video data. It offers the power of security from passive monitoring to intelligent analysis systems. The data obtained can be optimized for the management of daily life activities and for traffic analysis. The benefits of secured video streaming are utilized for remote access and third party integration and to implement security policies in design and implementation of the IoT network video system.

Therefore, the current smart camera suppliers are equipped with advanced features, bug fixes, and security patches.

1.10 IoT Security Activities

1.10.1 Device Manipulation

A device manipulation attack threatens the configuration, control, authenticity, and monitoring of IoT devices. Certain devices need timely updation. The update time of a device is prone to cause device failures or might also increase the system downtime. Thus, it is essential that the devices be reconfigured and updated in such a way that the revenue of the IoT network does not get affected.

Device manipulation looks after the secured establishment of the device identity in such a way that the device can be trusted. Thus, the goal of device manipulation is to monitor:

- Authentication
- Service provisioning
- Configuration and version control
- Maintenance of software updates.

1.10.2 Risk Management

Risk is a dynamic issue which concerns not only the vulnerabilities but also the impact of a threat on economy, privacy, and growth of the network. The risk assessment and a pre-planned strategy for risk avoidance are important, so that the legal compliances of companies, their business processes, standards, and infrastructure do not get shattered upon an unexpected external event.

1.10.2.1 Elements under risk management

The factors studied under risk management are:

1. Vulnerability: It is an application, a service, a configuration, or a hardware device of an IoT which can be exploited by an attacker, and is threat prone. Lack of computing power, inefficient encryption algorithms, etc., are vulnerabilities of a system.
2. Intent: Attackers conduct attacks to attain social, financial, or political benefits. The impact of the attack is as per their desires. The intent reflects the terror-oriented and malicious motives of the attacker.

3. Consequences: It is a tradeoff between the level of exploitation an attack can cause and the ability of a system to cope up with its impact. Certain attacks are intended for a prank but certain can cause huge economical harm as well as may lead to loss of lives.

1.10.2.2 Steps for risk management

Authentication and encryption mechanisms are used for risk management. Weak authentication opens the door to the outside world for attacks on the network. It is easy to obtain the login credentials and to masquerade the false identity. Inefficient encryption algorithms can be breached, and thus it is needed that encryption algorithms are strong enough to be computationally infeasible to be hacked.

1.10.2.3 Loopholes of current risk management techniques

1. In the world of cyber security, weak authentication is still a major problem that persists.
2. Passwords can be hacked and cracked in seconds.
3. Additional computing resources and memory are required for the implementation of strong encryption algorithms.
4. Key management is a difficult and complex task.

1.10.2.4 IoT risk management for data and privacy

The amount of data increases exponentially with IoT, and thus it raises the space requirements for transferring and storing these data. Data protection is although a burden task, but it is essential for business policies and decisions. The Iota network when scaled up has a higher degree of risks. Iota risk management for data concerns for segregating the individual and aggregate data, the important and unimportant data.

Iota risk management helps to determine the unacceptable risk conditions and their intensity of impact on safety and privacy of IoT network users.

1.11 Machine Learning in IoT

Machine Learning is the key to almost all the recent technologies developing nowadays. It is basically based on the concept of making use of all data that are collected by the machine for analysis. This data source ranges from entirely raw data to more processed information and also varies in size ranging till multiple terabytes. However, the implementation of such an advanced machine learning system specifically for IoT Security is not an easy task. It requires an amalgamation of fast processors, efficient classification algorithms, and most importantly effective decision making process based on the statistics. There is a phenomenal growth in IoT deployments around the globe, which is attracting the focus on IoT Security, exactly as the Cyber Security was followed by the growth of Internet in last few decades. Most of the technologists, researchers, and practitioners believe that securing IoT systems would be the prime most concern in coming years, which is to be dealt intelligently through machine learning approaches.

This situation is even more difficult to be programmed as IoT introduces a more number of treats than the Internet. IoT technologies are more exposed to unauthenticated intermediate message processors, open WiFi, multiple protocols, and spoofed sensors. Any vulnerable or compromised device in an IoT system is more dangerous than an intruder from outside and hence increases the probability of attacks. Also, all the devices of the IoT system are having their own memory and processing power, which enables them to bypass the control or to change it according to the intruder intentions. This also generates the possibility of newer treats to the system.

1.11.1 Need

A computer system can be secured by using latest security software, which works fine until the system is not connected to the Internet. Connecting to Internet invites much vulnerability in the system which requires updated security mechanisms to work continuously in order to keep the system secure. Many sophisticated software patches are available to achieve the high-end security, but this also requires enough memory and computing power at the host end. Unfortunately, in most of the IoT systems, the devices are having low computing power and minimum memory to accommodate such massive security mechanisms. This is the most important aspect which makes the IoT systems more vulnerable to security threats. Also, having the Internet access to the IoT system makes the situation even worse. Search engines like Shodan are the perfect example by which the openness of IoT systems may be visualized. And anything which is visible on Internet has cent percent chances to get hacked. Hence, to prevent such a hostile IoT environment, machine learning can be helpful to have a detailed analysis. IoT devices generated millions of data which could be a good source for machine learning approaches to have an estimation of abnormal activities and potential threats.

1.11.2 Levels of IoT Security

The IoT system consists of multiple protocols and various peer-to-peer communications among the devices involved. This multi-variant layers of operating systems require an effective security mechanism which should provide a comprehensive security solution while performing well on all the key points of the IoT system. Securing an IoT system requires implementation of security subroutines mainly at the following four levels.

1.11.2.1 Device

It is directly related with all the hardware and respective drivers associated with the devices which are the part of the IoT system deployed. It introduces the security at the physical layers of systems by implementing device authentication through MAC addresses and encryption keys, secure booting, and identification of devices.

1.11.2.2 Communication

It refers to the concept of securing the communication channels among the devices connected through an IoT system. Most of the communication channels are wireless, and hence the potential threats of attacking on these channels are very high. Sophisticated mechanisms like advanced public encryption, firewalls, web socket, virtual tunneling, and Secure WiFi are used to secure these open communication links. Also, due to often communication delays, these security mechanisms should be fast enough to cater the needs within a stipulated time frame.

1.11.2.3 Cloud

It refers to the backbone of IoT system, where all the data are collected, classified, analyzed, processed and then routed back in the system. It is the main software which is responsible for meeting the objectives of overall IoT system. Securing this part of IoT system is the most complex task, as most the security breaches are targeted here.

1.11.2.4 Life cycle

This is somewhat more comprehensive approach in the IoT system to provide security to the overall system while managing the system prompt and updating all the time. It ensures that the security mechanisms placed at all the fore-mentioned points should work cohesively with each other to build a higher level security protection. Various merits like risks analysis, auditing,

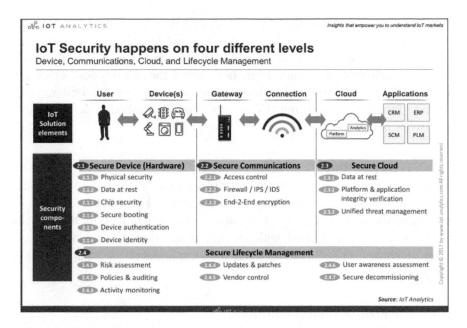

and activity assessment are regularly monitored to provide quick and correct results by the IoT system.

1.11.3 Automation of Security Mechanisms

With the growth of IoT systems, the requirement of security mechanisms has also increased. Generally, these activities are handled manually, like allotting and revoking of certificates, blacklisting of malicious nodes, and isolating the compromised devices. Better security systems are needed to be incorporated in IoT systems, which should not only be efficient but intelligent enough to minimize the human intervention for taking small decisions. For example, simple classification of malicious nodes as per their behavior can be taken care by an intelligent subroutine which can learn by historic data. This automation will enable the IoT system to take appropriate security decision while making it autonomous. There are many ways through which we can achieve automation, like we may create an inference system, or a neural network, or combination of any machine learning techniques. Machine learning techniques can be classified as (1) supervised machine learning, where predictions are made as per a given set of samples by searching for patterns within the labels assigned to data points, (2) unsupervised machine learning, where there are no labels associated with data points. It initially organizes the data into a group of clusters to describe its structure for further analysis, and (3) reinforcement machine learning, where an action is to be taken first, and later learn how good the action was. Over the time, it changes the strategy to learn better and achieve the best reward. Machine learning for securing IoT systems can be applied at device points or network points.

1.11.3.1 Device-based solutions

The major concern in device-based security solutions is the low memory and less storage capacity for executing the subroutines. For a better analysis of threats and the record keeping of signatures and authentication, the devices are required to be equipped with enough processing power and storage. Various techniques like threading can help to implement high-level protection in less resources.

1.11.3.2 Network-based solutions

Securing IoT systems can also be done at the network level by registering all the devices to the network and regular auditing of data traffic to and from the IoT system. And if anything goes beyond expected or suspicious, alarms can be activated for safe guarding the data and control points of the system. This traffic monitoring scheme could be used to identify and classify the compromised nodes based on their behavior and past experiences.

1.11.4 Classification of IoT Security Techniques

1.11.4.1 Network security

Protecting and securing the IoT network is the key for the high-end security mechanism. However, it is often a more complex task due to the heterogeneous communication protocols, standards, and most importantly device capacities. Firewalls and intrusion prevention methods can be applied on network to avoid the potential threats from intruders.

1.11.4.2 Authentication

It enables the users to authenticate the trusted IoT devices by using pin codes, certificates, or biometrics. It is a manual but yet a powerful technique to avoid attacks.

1.11.4.3 Encryption

It is always better to encrypt the data and control to achieve better security. Various encryption protocols may be used to keep the data safe from outsiders. Public or private key encryption may be used secure the IoT system.

1.11.4.4 Analytics

This involves the collection, organization, and analysis of the data on the IoT system and generation of alerts if any activity falls on the suspicious category. New and better machine learning methods can be deployed to perform such analysis.

1.12 Conclusions

The fourth industrial revolution has opened a broad pathway for machine-to-machine (M2M) communication to facilitate the extreme automation. Now, machines can communicate and share information to understand each other for specific objectives. These objectives are required to be well programmed to maintain the privacy and to be well equipped for possible security breaches. However, many times due to the minimum human intervention in this whole scenario, the security is considerably compromised. Therefore, advanced techniques need to be implemented at negotiation, authentication, execution, and information exchange points in M2M communication. These technologies are now becoming intelligent to sense and detect the security attacks while improving the overall system and making it more resilient against modern IoT attacks. This situation is more vulnerable because of version incompatibility issues, and versioning of machines and firmwares. The presented chapter discussed the conceptual framework for these technologies and related issues.

2

Internet of Things Privacy, Security, and Governance*

Gianmarco Baldini[1], Trevor Peirce[2], Marcus Handte[3],
Domenico Rotondi[4], Sergio Gusmeroli[4], Salvatore Piccione[4],
Bertrand Copigneaux[5], Franck Le Gall[5], Foued Melakessou[6],
Philippe Smadja[7], Alexandru Serbanati[8], and Julinda Stefa[8]

[1]Joint Research Centre – European Commission, Italy
[2]AVANTA Global SPRL, Belgium
[3]Universität Duisburg-Essen, Germany
[4]TXT e-solutions S.p.A., Italy
[5]Inno TSD, France
[6]University of Luxemburg, Luxemburg
[7]Gemalto, France
[8]Sapienza University of Rome, Italy

2.1 Introduction

Internet of Things (IoT) is a broad term, which indicates the concept that increasingly pervasive connected devices (embedded within, attached, to or related to "Things") will support various applications to enhance the awareness and the capabilities of users. For example, users will be able to interact with home automation systems to remotely control the heating or the alarm system.

The possibility of implementing "intelligence" in these pervasive systems and applications has also suggested the definition of "Smart" contexts, where digital and real-world objects cooperate in a cognitive and autonomic

---

*This chapter is reproduced content from the published article in Internet of Things: Converging Technologies for Smart Environments and Integrated Ecosystems, 207–224, 2013.

way to fulfill specific goals in a more efficient way than basic systems implemented on static rules and logic. While full cognitive and autonomic systems may still be years away, there are many automated processes and automated Internet process which we take for granted every day. So why should the Internet of Things (IoT) require special attention when it comes to privacy, security, and governance? Does not the established Internet have these matters dealt with sufficiently already, given that through just about every smartphone anywhere there are already a wide variety of sensors capturing information which we share on the Internet, e.g., photos, videos, etc.? Why is IoT any different?

First, IoT is different because it will be possible and likely that objects will autonomously manage their connections with the Internet or, this will be done upon the request of someone or something remotely. When someone shares a video or a photo taken on their mobile phone over the Internet, they "call the shots." With IoT, potentially someone else is in charge. For reasons largely similar to this, the topics of privacy, security, and governance are very important if not vital to the success of IoT in order to establish and maintain stakeholder trust and confidence. Yes, there is a large overlap between IoT and Internet in many areas pertaining to trust; however, IoT brings many new specific dimensions too.

The adoption of IoT essentially depends upon trust. Moreover this trust must be established and maintained with respect to a broad group of stakeholders; otherwise, IoT will face, to some degree or other, challenges which may restrict adoption scope or delay its timing. Note that with social media you make the conscious choice to publish; some IoT applications may adopt the same or similar model but there may be other instances or applications where this will not be the case. This remote control is not essentially bad. For example, if you were incapacitated due to an accident, it could be advantageous that rescue services would be able to access objects in your environment to locate you or communicate with you. However, if these devices were configured to automatically inform your children what presentations had been bought or not bought, this could spoil much of the excitement of receiving gifts.

Facebook's withdrawn Beacon[1] service was accused of this when shoppers' purchases were automatically published online resulting in a public outcry and class-action in the US post-holidays (Christmas). There are also potential ethical issues if essential services oblige you to use IoT-connected

---

[1] http://en.wikipedia.org/wiki/Facebook_Beacon

health monitoring devices. Also a number of Internet services are already struggling with the ethical issues of capturing and publishing information affecting third parties where appropriate permissions have not been sought from the third parties involved, e.g., Street View[2]. Trust, privacy, and governance aspects of IoT rely for the most part upon security [1]. Security in its broadest definitions includes health and well-being as well as other forms of protection. These aspects need to be viewed from the perspectives of the majority if not all the principle stake-holder groups and extended to include the relevant influencing and influenced elements of the general environment. Today from the European Commission's perspective, the essential focus for security is the protection of health and the avoidance of potential super-power control being established by enterprises. The objectives are not currently focused upon seeking specific IoT measures to deter cybercrime, cyberwarfare, or terrorism. Without sufficient IoT security, it is highly likely that some applications will more resemble the Intranet of Things rather than the IoT [2] as users seek to place their own proprietary protection barriers and thus frustrating broad interoperability. Many of the device connections to the Internet today more closely resemble the Intranet of Things which differs dramatically from the vision for the IoT, the latter being a much more open and interoperable environment allowing in theory the connection with many more objects and with their multiple IoT compatible devices.

The future of IoT is not only influenced by users. The potential autonomy of IoT or lack of control over IoT by those it impacts will doubtless generate IoT adoption resistance potentially manifested by public protests, negative publicity campaigns, and actions by governments. Indeed, many IoT foundation technologies have been influenced during the last 10 years by the developing concerns which have been labeled as "threats to privacy." Privacy itself is multidimensional. Popular definitions focus upon individual freedoms, or the "right to be left alone." In reality, privacy encompasses the interests of individuals, informal groups, and including all forms of organizations and is therefore a complex multidimensional subject.

In an age of social media, it is interesting to see growing examples of how industry groups and governments begin to encourage greater individual responsibility for protecting our own privacy, defending our virtual representation in order to protect our identity and diminish the challenges of real-world or virtual-world authentications and authorization processes. Through IoT, this may become an increasingly "hard sell" as individuals

---

[2]http://en.wikipedia.org/wiki/Google_Street_View

begin to realize that any efforts individuals take to protect their own identities have almost no influence due to the amounts of information smart objects are collecting and publishing on the Internet. Ideally, IoT would provision for flexibility enabling it to be suitably synchronized with the evolution of the development and use of the wider Internet and the general real-world environment.

One specific challenge in IoT is the control of the information collected and distributed by mobile devices which are increasingly small and pervasive like RFID or future micro–nano sensors, which can be worn or distributed in the environment. In most cases, such devices have the capability of being wireless connected and accessible. In this context, the challenge is to ensure that the information collected and stored by micro/nano-RFID and micro/nano sensors should be visible only to authorized users (e.g., the owner or user of the object); otherwise, there could be a breach of security or privacy. For example, the owner of a luxury good may not want anybody to know that the luxury good is in a suitcase. The watch in the suitcase may be hidden from view, but it can be easily tracked and identified through wireless communication. In a similar way, the information collected by the body sensors applied to an elderly person should not be accessible by other persons apart from the doctor. Access control mechanisms for these wireless devices should be implemented and deployed in the market, but security and privacy solutions are not easy to implement in micro–nano devices because of the limitations in computing power and storage. At the same time, security and privacy should not hamper business development of micro–nano technologies. Keys' management and deployment can also be complex to implement. Tradeoffs should be identified and described. These are goals for research activity.

One aspect which often gets overlooked particularly frequently by those of us who entered adulthood before the year 1990 is the importance of the virtual-world. Today, the virtual identities of children are as important to them if not more so than their real-world identities. Within the virtual-world, there exists most if not all of the things we find in the real-world including objects, machines, money, etc. IoT includes the real- and virtual-worlds, and indeed, it is capable of establishing an important bridge between the two. This bridge is likely to grow and become more relevant in the life of citizens in the future. New devices like Google Glass or future Intelligent Transportation Systems' (ITSs) applications in cars will propose "augmented reality" where the integration of digital and real-word information is used to compose sophisticated applications. This trend highlights even more the need for security and privacy, because data breaches in the virtual-world can have consequences in the real-world. In some contexts and applications, security and privacy threats can even become safety threats with more dramatic consequences for the lives of the citizen. As a conceptual example, actuators in the real-world may be set remotely within a "smart house" to provoke fires or flooding.

2.2 Overview of Activity Chain 05 – Governance, Privacy, and Security Issues

The European Research Cluster on the IoT has created a number of activity chains to favor close cooperation between the projects addressing IoT topics and to form an arena for exchange of ideas and open dialog on important research challenges. The activity chains are defined as work streams that group together partners or specific participants from partners around well-defined technical activities that will result into at least one output or delivery that will be used in addressing the IERC objectives. IERC Activity Chain 05 is a cross-project activity focused on making a valued contribution to IoT privacy, security, and governance among the EC funded research projects in the area of IoT. As described in [3], the three aspects are closely interlinked "Privacy, security, and competition have been identified as the main issues related to IOT Governance; however, those issues should not be discussed in a separate or isolated way" [3]. In the same reference, it was also highlighted the challenge to define a common agreed definition for Governance of IoT. In a similar way, the concepts of security and privacy do not have a uniform definition in the literature even if there is a common agreement on these concepts. Overall, the main objective of the Activity Chain 05 is to identify research challenges and topics, which could make IoT more secure for users (i.e., citizen, business, and government), to guarantee the privacy of users and support the confident, successful, and trusted development of the IoT market. In comparison to IoT initiatives in Europe or at a global level (e.g., IGF), Activity Chain 05 does not define government policies but focuses upon research (which could eventually be used to support policies or standardization activities). The following sections provide an overview of some contributions which European Commission funded projects associated with Activity Chain 05 have made to IoT privacy, security, and governance.

2.3 Contribution from FP7 Projects

2.3.1 FP7 iCore Access Framework (iCore Contribution)

The iCore cognitive framework is based on the principle that any real-world object and any digital object that is available, accessible, observable, or controllable can have a virtual representation in the "Internet of Things," which is called virtual object (VO). The VOs are primarily targeted to the abstraction of technological heterogeneity and include semantic description of functionality that enables situation-aware selection and use of objects. Composite virtual objects (CVOs) use the services of VOs. A CVO is a cognitive mash-up of semantically interoperable VOs that render services in accordance with the user/stakeholder perspectives and the application requirements.

The overall layered approach of the iCore project is provided in Figure 2.1. The first cognitive management layer (VO level cognitive framework) is responsible for managing the VOs throughout their lifecycle, ensuring reliability of the link to the real-world object/entity (e.g., sensors, actuators, devices, etc.). They represent, for example, in a logistic related scenario, tracking temperature-controlled goods' transport, individual goods' boxes are represented by VOs and the container transported by a truck is a VO as is the truck itself. IoT-related applications can interface for different service reasons each of these VOs separately.

The second cognitive management layer (CVO level cognitive framework) is responsible for composing the VOs in CVO. CVOs will be using the services of VO to compose more sophisticated objects. In our example, the combination of the truck and the transported goods is represented in the cognitive framework as a CVO.

The third level (user level cognitive framework) is responsible for interaction with users/stakeholders. The cognitive management frameworks will record the users' needs and requirements (e.g., human intentions) by collecting and analyzing the user profiles and stakeholders' contracts (e.g., service level agreements) and will create/activate relevant VO/CVOs on behalf of the users.

2.3.2 IoT@Work Capability-based Access Control System (IoT@Work Contribution)

The IoT envisages new security challenges, including in the area of access control that can hardly be met by existing security solutions. Indeed, IoT is a more demanding environment in terms of scalability and manageability

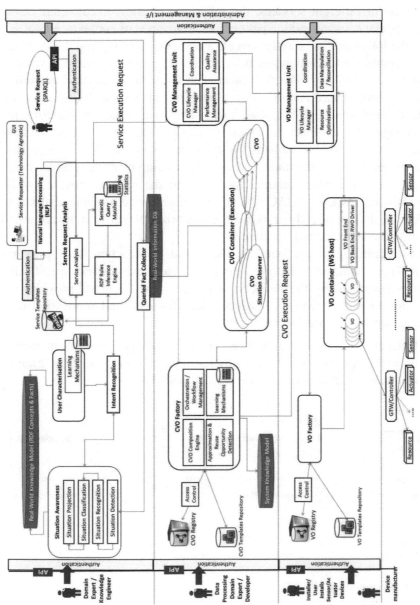

Figure 2.1 iCore framework.

due both to the potentially unbounded number of things (resources and subjects), the expected most relevant need to support the orchestration and integration of different services, the relevance of short-lived, often casual and/or spontaneous interaction patterns, the relevance of contexts, etc.

In the following, we shortly provide a description of the capability-based access control (in the following, referred to as CapBAC) system developed within the EU FP7 IoT@Work project. The CapBAC is devised according to the capability-based authorization model in which a capability is a communicable, unforgeable token of authority. This token uniquely identifies the granted right(s), the object on which the right(s) can be exercised, and the subject that can exercise it/them. As depicted in Figure 2.2, a capability-based system reverses the traditional approach being now the user in charge of presenting his/her/its authorization token to the service provider, while in a traditional ACL or RBAC system, it is the service provider that has to check if the user is, directly or indirectly, authorized to perform the requested operation on the requested resource.

The CapBAC system borrows ideas and approaches from previous works [4] extending and adapting them to IoT requirements and, specifically, the ones envisaged by the IoT@Work project. The CapBAC provides the following additional features that constitute the essential innovation over previous capability-based techniques:

(a) Delegation support: A subject can grant access rights to another subject, as well as grant the right to further delegate all or part of the granted rights. The delegation depth can be controlled at each stage.

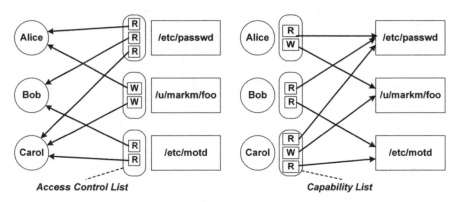

Figure 2.2   ACL vs. capability-based authorization models.

(b) Capability revocation: Capabilities can be revoked by properly authorized subjects, therefore solving one of the issues of capability-based
approaches in distributed environments.

(c) Information granularity: The service provider can refine its behavior and
the data it has to provide according to what is stated in the capability
token. Figure 2.3 exemplifies the usage of a capability-based access
control approach to manage a simple situation: Bob has to go on holidays
and his house needs some housekeeping while he is away. Dave offered
to take care of Bob's house for his holiday's period. Bob provides to
Dave an access token that: (a) identifies that Dave has the only subject
entitled to use the token, (b) states what Dave can actually perform, and
(c) states for how many days Dave can do these actions.

Bob and Dave do not need to establish trust relationships among their
authentication and authorization systems. Bob's house appliance recognizes
the access token created by Bob and Dave has only to prove that he is the
subject (grantee) identified by the capability token as entitled to do specific
housekeeping activities for the holidays' period. The above mechanism is
very intuitive, easy to understand, and easy to use. CapBAC is well suited to

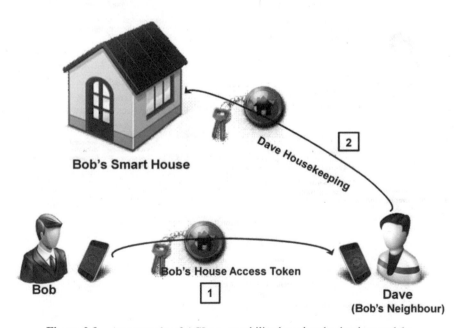

Figure 2.3   An example of ACL vs. capability-based authorization models.

manufacturing contexts where there are many subjects, internal (e.g., workers and production supervisors) and external (e.g., suppliers and maintainers), that need access both directly (e.g., via mobile or desktop computing sets) or indirectly (e.g., via application services) to devices, data, and services in the manufacturing plant. Most of, if not all, these elements require enforcement of strictly access control policies and finer-graded access control, and, at the same time, a management effort that has to be decoupled from the number of managed resources or subjects, especially when many subjects are external ones.

The CapBAC architectural elements can be shortly characterized as follows (Figure 2.4):

- The resource object of the capability (Service A in Figure 2.4); it can be a specific data or device, a service, or any accessible element that can be univocally identified and/or actable on (like resource);
- The authorization capability that details the granted rights (and which ones can be delegated and, in case, their delegation depths), the resource on which those rights can be exercised, the grantee's identity, as well as

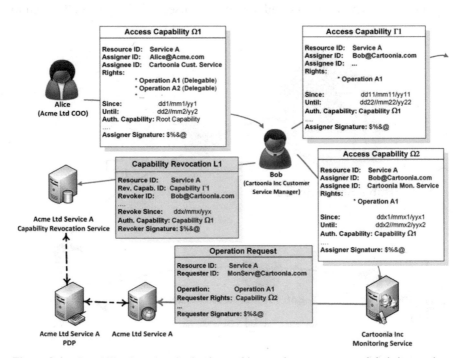

Figure 2.4 Capability-based authorization architectural components and their interactions.

additional information (e.g., capability validity period, XACML conditions, etc.). An authorization capability is valid as specified within the capability itself or until it is explicitly revoked;

- The capability revocation is used to revoke one or more capabilities. Like a capability, a capability revocation is a communicable object a subject, having specific rights (e.g., the revoker must be an ancestor in the delegation path of the revoked capability), creates to inform the service in charge of managing the resource that specific capabilities have to be considered no more valid;
- The service/operation request is the service request as envisaged by the provided service with the only additional characteristics to refer or include, in an unforgeable way, a capability. For example, for a RESTful service, an HTTP GET request on one of the exposed REST resources has to simply include the capability and its proof of ownership to use our access control mechanism;
- The PDP (policy decision point) is a resource-agnostic service in charge of managing resource access request validation and decision. In the CapBAC environment, it deals with the validation of the access rights granted in the capability against local policies and checking the revocation status of the capabilities in the delegation chain;
- The resource manager is the service that manages service/access requests for/to the identified resource. The resource manager checks the acceptability of the capability token shipped with the service request as well as the validity and congruence of the requested service/operation against the presented capability. It acts as an XACML Policy Enforcement Point (PEP) which considers the validation result of the PDP;
- The revocation service is in charge of managing capability revocations.

2.3.3 GAMBAS Adaptive Middleware (GAMBAS Contribution)

The GAMBAS project develops an innovative and adaptive middleware to enable the privacy-preserving and automated utilization of behavior-driven services that adapt autonomously to the context of users. In contrast to today's mobile information access, which is primarily realized by on-demand searches via mobile browsers or via mobile apps, the middleware envisioned by GAMBAS will enable proactive access to the right information at the right point in time. As a result, the context-aware automation enabled by the GAMBAS middleware will create a seamless and less distractive experience for its users while reducing the complexity of application development.

As indicated in Figure 2.5, the core innovations realized by GAMBAS are the development of models and infrastructures to support the interoperable representation and scalable processing of context, the development of a generic, yet resource-efficient framework to enable the multimodal recognition of the user's context, protocols, and mechanisms to enforce the user's privacy as well as user interface concepts to optimize the interaction with behavior-driven services.

From a security and privacy perspective, the developments in GAMBAS are centered on a secure distributed architecture in which data acquisition, data storage, and data processing are tightly controlled by the user. Thereby, security and privacy are based on the following elements.

- Personal acquisition and local storage: The primary means of data acquisition in GAMBAS are personal Internet-connected objects that are owned by a particular user such as a user's mobile phone, tablet, laptop, etc. The data acquired through the built-in sensors of these devices are stored locally such that the user remains in full control. Thereby, it is noteworthy that the middleware provides mechanisms to disable particular subsets of sensors in order to prevent the accumulation of data that a user may not want to collect and store at all.

- Anonymized data discovery: In order to enable the sharing of data among the devices of a single user or a group of users, the data storages on the local device can be connected to form a distributed data processing system. To enable this, the GAMBAS middleware introduces a data discovery system that makes use of pseudonyms to avoid revealing the user's identity. The pseudonyms can be synchronized in an automated fashion with a user-defined group of legitimate persons such that it is possible to dynamically change them.

- Policy-based access control: To limit the access to the user's data, the networked data storages perform access control based on a policy that can be defined by a user. In order to reduce the configuration effort, the GAMBAS middleware encompasses a policy generator tool that can be used to derive the initial settings based on the user's sharing behavior that he exhibits when using social services.

- Secure distributed query processing: On top of the resulting set of connected and access-controlled local data storages, the GAMBAS middleware enables distributed query processing in a secured manner. Toward this end, the query processing engine makes use of authentication mechanisms and encryption protocols that are bootstrapped by means of novel key exchange mechanisms that leverage the existing web-infrastructure that is already used by the users.

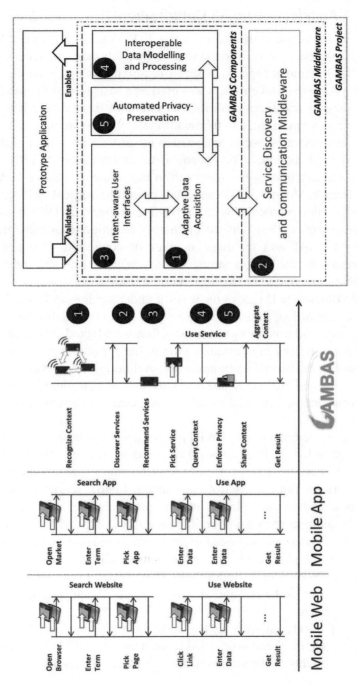

Figure 2.5 GAMBAS middleware.

2.3.4 IoT-A Architecture (IoT-A Contribution)

Security is an important cornerstone for the IoT. This is why, in the IoT-A project, we deemed as very important to thoroughly address security and privacy issues in various aspects. A set of requirements based on the input of external and internal stakeholders was used as a basis for the identification of the mechanisms and functionalities that guarantee user data privacy and integrity, user authentication, and trustworthiness of the system.

These functionalities were analyzed and orchestrated in Functional Groups (FG) and Functional Components (FC) in the frame of WP1. High-level PS&T specifications were integrated in the frame of the IoT-A Architectural Reference Model (ARM) and then passed to vertical WPs dealing with communication protocols (WP3), infrastructure services (WP4), as well as hardware aspects (WP5). Due to the highly heterogeneous environment provided by the IoT and the huge number of connected, (autonomous) devices foreseen by analysts, a strong focus was placed on scalability and interoperability.

The ARM document [5] paves the way for understanding and adopting the open architecture of IoT-A, as well as provides the overall definition of IoT security, privacy, and trust design strategies that we adopted. Then, in WP3, we analyzed the security of communication in the peripheral part of the IoT and its impact on the overall communication architecture. In this context, we investigated HIP and HIP-BEX protocols, as well as considered issues like mobility, collaborative key establishment, and securing network entry with PANA/EAP.

Then, within the framework of WP4 [6], we developed a secure resolution infrastructure for IoT-A. It ensures privacy and security for the resolution functions and offers the basis for other security functionalities outside the resolution infrastructure. It controls the access to IoT resources, to real-world entities, and to the related information including their respective identifiers. In addition, the resolution infrastructure also provides support for pseudonymity: A user does not need to reveal his/her identity when using an IoT resource or a higher level service. To achieve all this, various security components were developed (Figure 2.6). They deal with authorization and authentication, key exchange and management, trust and reputation, and identity management.

Finally, WP5 deals with privacy and security at a device level. In particular, it describes the mechanisms needed to authenticate RFID devices and to provide confidentiality of the communication between the reader and the tag.

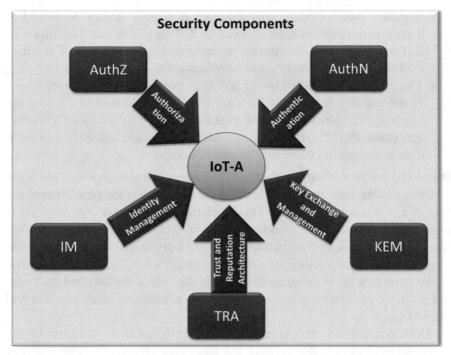

Figure 2.6   Components for privacy and security in the IoT-A resolution infrastructure.

The PS&T features of the IoT-A architecture will be tested in the forthcoming IoT-A eHealth use case.

2.3.5 Governance, Security, and Privacy in the Butler Project (Butler Contribution)

The goal of the BUTLER project is the creation of an experimental technical platform to support the development of the IoT. The main specificity of the BUTLER approach is its targeted "horizontality": The vision behind BUTLER is that of a ubiquitous IoT, affecting several domains of our lives (health, energy, transports, cities, homes, shopping, and business) all at once. The BUTLER platform must therefore be able to support different "Smart" domains, by providing them with communication, location, and context awareness abilities, while guaranteeing their security and the privacy of the end users. The issue of security and privacy is therefore central in the BUT-LER project and develops in several requirements, the main requirements relate to:

- Standard issues of data security, both at a data storage level and at a data communication level, exist in IoT applications. The diversity and multiplicity of the "things" connected by the IoT and of the data exchanged further amplify and complicate these requirements.
- The application enabled by the IoT may pose additional privacy issues in the use that is made of the data from the collection of data by the applications (which should be conditioned by an "informed consent" agreement from the user), to the profiling, exchange, and sharing of these data necessary to enable true "context awareness."

Data technical protection[3] mechanisms include two major aspects. One is the protection of the data at data storage and the other one is the protection of the data at a communication level. The protection of data at a communication level is one of the major areas of research. Many communication protocols implement a high level of end-to-end security including authentication, integrity, and confidentiality. At a communication level, the major issue is the deployment process of the security keys and the cost of the required hardware and software environment to run the security algorithms in an efficient and secured way.

However, privacy and security do not only refer to security of the exchange of data over the network but shall also include: (a) Protection of the accuracy of the data exchanged, (b) protection of the server information, (c) protection of the usage of the data by explicit, dynamic authorization mechanisms, (d) selected disclosure of data, and (e) the implementation of "transparency of data usage" policies.

The BUTLER project also addresses the security and privacy challenges from the point of view of their implication on business models. To specify the horizontal IoT platform envisioned in BUTLER, the project started from the gathering and analysis of the requirements from up to 70 use cases. The analysis of these use cases not only produced requirements for the specification of the platform but also valuable information on the potential socio-economic impact of the deployment of a horizontal IoT and on the impact on the associated business models.

If treated accordingly, the ethics and privacy issues transform from a threat to an opportunity. A better understanding of the service by the user increases acceptance and creates trust in the service. This trust becomes a

---

[3]An exhaustive study of the security enabling technologies is available in "D2.1 Requirements, Specification and Security Technologies for IoT Context-aware network." http://www.iot-butler.eu/download/deliverables

competitive advantage for the service provider that can become a corner stone of his business model. In turn, the economic interest of the service providers for ethics and privacy issues, derived from this competitive advantage, becomes a guarantee for the user that his privacy will be respected. The BUTLER project research on the implication of the ethics, privacy, and data security on the business models and socio-economic impact will be published in Deliverable 1.4 (May 2013) and Deliverable 1.3 (September 2014).

The involvement of end users in proof of concepts and field trials is another specificity of the BUTLER project. The end user involvement is key not only to validate the technical qualities of the BUTLER platform (technology feasibility, integration and scaling) but also to assess the perception of end user and their acceptance of the scenario envisioned for the future "horizontal" IoT.

However, the involvement of end users in the scope of the project requires handling their data and privacy concerns carefully. The detailed specification of the field trials and proof of concept are described in Deliverable 1.2 (scheduled for end of May 2013). The following issues must be considered in the organization of end user involvement:

(a) Technical security mechanisms must be set up to ensure the security and privacy of the participants. This involves secured data communication and storage, and in the scope of the BUTLER project, it is addressed by the enabling security technologies developed and integrated in the BUTLER platform.
(b) The participants must be well informed of the scope and goal of the experiment. In the case of BUTLER, this involves specific efforts to explain the scope and goal of the project to a larger public.
(c) The consent of the participants must be gathered based on the information communicated to them. The consent acknowledgment form must remind the participants of their possibility to refuse or withdraw without any negative impact for them.
(d) finally both a feedback collection and a specific complaint process have been designed to offer the possibility to the participants to raise any issue identified.

2.4 Conclusions

IoT applications and supporting stakeholders can all mutually benefit from the establishment of a trusted IoT. Trust means establishing suitable provisions for privacy, security, and governance. To put in place and maintain trust means fulfilling today's needs while providing sufficient future provisions to meet naturally evolving stakeholder requirements and expectations. Consensus necessary for the formation of successful standards and guidelines can come only through dialogs. Activity Chain 05 provides such a platform for information exchange and mutual understanding as well as in providing valued leadership. The research projects within Activity Chain 05 all contribute to advancing IoT adoption, some having a universal IoT application value while others provide significant enhancements to specific IoT application groups. Making this landscape clearer, identifying the gaps for further research as IoT develops and, assisting the progression of research toward standardization and adoption remain the principle challenges for Activity Chain 05. Another role for Activity Chain 05 is raising awareness and promoting adequate consideration of IoT privacy, security, and governance within the other activity chains of the IERC and the wider stakeholder community.

References

[1] Roman, R., Najera, P., Lopez, J., Securing the Internet of Things. *Computer*, 44, 51–58, 2011.

[2] Zorzi, M., Gluhak, A., Lange, S., Bassi, A., "From today's INTRAnet of things to a future INTERnet of things: a wireless- and mobility-related view," in *IEEE Wireless Communications*, 17, 44–51, 2010.

[3] *Final Report of the EU IOT Task Force on IOT Governance.* Brussels, 2012.

[4] Gusmeroli, S., Piccione, S., Rotondi, D., "IoT Access Control Issues: A Capability Based Approach," in *Proceedings of 6th Int. Conf. on Innovative Mobile and Internet Services in Ubiquitous Computing (IMIS-2012)*, 787–792, 2012.

[5] Carrez, F. (ed.), *Converged architectural reference model for the IoT*, available at: http://www.iot-a.eu/public/public-documents [Accessed 10 May 2013].

[6] Gruschka, N., Gessner, D. (eds.), *Concepts and Solutions for Privacy and Security in the Resolution Infrastructure*, available at: http://www.iot-a.eu/public/public-documents [Accessed 10 May 2013].

3

IoT Data Processing: The Different Archetypes and Their Security and Privacy Assessment

Pijush Kanti Dutta Pramanik and Prasenjit Choudhury

Dept. of Computer Science & Engineering, National Institute of Technology, Durgapur, India

Undoubtedly, in recent years, Internet of Things (IoT) has created the maximum ripples in the IT industry. Large-scale implementation of IoT will generate a massive amount of data. To obtain actionable knowledge, these data must be processed. But the complex nature of these data and the IoT architecture as well have made the processing of a complicated task. This chapter looks into the fact how it becomes challenging to the computing fraternity to process the diverse types of dynamic data generated from diverse heterogeneous IoT devices. This chapter will help readers in having an overview of the different architectures for IoT data processing. Each architecture is discussed decoratively along with the associated challenges. The open and ubiquitous nature of IoT makes it more vulnerable to the security threats. The issues pertaining to security and vulnerabilities that are faced by the architectures are discussed specifically. Along with a general discussion on the security and privacy issues in IoT, the authors have also portrayed a different view on the overhyped concerns for the security and privacy aspects in IoT. The purpose of this chapter is to make readers able to decide on the suitable data processing architecture for their IoT applications considering different factors such as cost, response time, etc., with a special focus on security and privacy.

3.1 Introduction

Internet of things (IoT) has brought probably the largest wave in the cyber industry after the Internet and mobile communications [1]. IoT refers to an interconnected setup [2] which allows remote accessing of the objects (things) over the Internet. These things are typically furnished with some sort of sensors and may include mobile devices, washing machines, street cameras, wearable devices, medical equipment, etc. As the number of such connected devices is growing at an express pace, by the next few years, it will become virtually boundless. This extensive number of things will result in an unforeseen amount of data generation. Besides others, the most vital but obvious concern is to be addressed that how to process such a huge and ever-increasing amount of data? Processing this humongous data is definitely challenging, especially, considering the heterogeneous and dynamic nature of IoT. To utilize the IoT systems most advantageously and effectively, an efficient data processing system is needed that can address different issues arises in the execution of IoT data capturing, processing, and consuming. Business organizations are recognizing great opportunity to mould their business to IoT-oriented. Understanding the nitty-gritty of different architectures will help the organizations in implementing IoT judiciously.

Since the IoT devices are typically characterized by a limited computational and storage capacity, it requires some external data processing means. One of the crucial decisions has to be taken by the IoT developers and implementers is that where to process the vast IoT data. Whether it is to be processed within the IoT devices or they should be processed in a remote centralized data center or in between somewhere else. Whether organizations should process on-premises using own infrastructure or a third-party service should be called on and so on.

Every networked system is vulnerable to security and privacy threats. IoT is also no exception. The widespread use of IoT on a huge scale has made the risk graver. The security of IoT data is of paramount importance, as it can open up vulnerabilities to both the user and the system due to the fragile nature of the data.

The rest of the chapter is organized as follows. In Section 3.2, some typical properties of IoT have been enlisted. In Section 3.3, we shall mention the challenges that are to be dealt with in IoT data processing. Section 3.4 addresses the core theme of this chapter, i.e., all the possible architectures that can be used for processing IoT data considering different factors. The security and privacy issues of each archetype are identified in Section 3.5.

A general discussion on the security and privacy issues in IoT has been stated in Section 3.6, whereas in Section 3.7, we present a rational view on whether the overstressed concern on the security and privacy in IoT is really that worrisome. Finally, Section 3.8 concludes the chapter.

3.2 Properties of IoT Data

As we mentioned, IoT will generate a massive volume of data. But since most IoT systems contain varieties of devices with diverse hardware, IoT data are not only voluminous, but they also exhibit the following characteristics:

- Heterogeneous
- Multi-dimensional
- Continuous with high velocity
- Dynamic and inconsistent
- Real-time and streaming
- Volatile and ad hoc
- Strong spatial and temporal dependency
- Varying data quality
- Diverse data structures and data types

3.3 Challenges in IoT Data Processing

The IoT is still in its adolescent stage, but considering the amount of data it generates, it is already imposing daunting challenges to the traditional data processing technologies. The specific properties mentioned in the previous section set additional challenges and make it difficult to process IoT data by using traditional data processing approaches and platforms. Some of the prominent challenges are:

- Limited or no processing power
- Communication restraints
- Limited memory size
- Real-time processing
- Handling high ingestion rate
- Preserving situation and context-awareness
- Energy constraint
- Supporting scalability
- Providing fault tolerance
- Ensuring security and privacy

3.4 IoT Data Processing Architectures

3.4.1 Grid Computing

Grid computing is a distributed system that allows seamless access to a computing grid made of a collection of computing resources connected through a network. Grid computing offers supercomputing like computing power utilizing intra- and/or inter-organizational computing resources such as desktops, clusters, RAIDs, etc. Grid computing is particularly suited for organizational IoT. It requires demanding computing infrastructure to process, store, and evaluate IoT data which surely raise the IT budget overwhelmingly. Instead of spending on third-party computing resources (e.g., the Cloud), organizations can make use of their existing IT infrastructure. In-house computing resources such as desktops and employees' personal portable computers including smartphones and tablets can be utilized to form a grid or pool of resources. IoT data should be gathered by a designated central entity that will assign, distribute, and schedule processing jobs to the potential computing sites. After completion, results are collected, assembled, and sent to concerned entities. Using grid computing to process IoT data entices many advantages such as:

- Better utilization of existing infrastructure, do not require additional setup.
- Less expensive compared to other options, do not have to pay out for external services.
- Self-sufficient, do not have to rely on third-party services.
- Just-in-time processing of IoT data, hence better response time.
- Communication cost and delay are avoided.

But the obvious drawback of grid computing is that the organizations have to bear the overhead of managing and maintaining the infrastructure, applications, and data. Hence, small and medium and also rookie organizations may not prefer this option.

3.4.2 Cloud Computing

Cloud computing has been the most favorable platform for IoT data processing. It provides centralized access to computing resources at a lower cost. Applications and services are hosted in the Cloud. IoT data are collected and sent to the Cloud for processing, storing, and analysis. Users and applications either subscribe for events prior or query to the Cloud for desired services. Powerful processors and massive storage are the lucrative options offered by

Cloud computing that IoT leverages. Thanks to Cloud's scalable services, IoT can collect data from any number of devices and store them indefinitely and most importantly securely. IoT can harness powerful Cloud-based applications and Big Data techniques, e.g., HBase and MapReduce for processing and analysis [3]. Cloud computing for IoT is widely accepted especially to those organizations who do not have sufficient existing IT infrastructure and also unwilling in upfront investment on this account.

Problem with Cloud Computing as Processing Architecture for IoT Data: One aspect of concern for which Cloud computing may not go well with IoT data is the fact that in most of the IoT systems, the operated data are real-time and the applications which consume these data are non-delay-tolerant. For example, disaster monitoring and control systems or industrial automation or smart trafficking all require real-time access to the sensor data. These applications cannot afford to carry the raw data to be processed on a remote site. In these cases, generated data should be processed at a site that is close at hand to the devices, so that the processing latency is minimized for immediate response. Another problem with Cloud computing is that all the IoT data are sent to the Cloud server for processing. This puts an immense burden on the underlying network. And the worst part is that the most of these data may not be needed to be processed at all. Actually, in the early phase of IoT evolution, application logic was embedded into the IoT devices' firmware and the processing was done at the device [4]. Though it allowed real-time processing and instantaneous response, it was hard to upgrade and reprogram those devices if any changes to the application had to be instigated [5]. And also, because of resource limitation, sophisticated algorithms could not be run on the IoT devices. Hence, it was recommended to take the application logic and processing out of the data-originating devices and delegate to a Cloud server [6]. So, now the entire raw IoT data are transported to and processed in the Cloud instead of the device. Since the data processing task has been moved out from the local network to a remote data center, this concept is called "*out-of-network processing*" [5]. But, as we have seen, this has introduced typical complications of network congestion and latency. To get rid of these problems, researchers pondered an intermediate approach and came out with an innovative solution as "*in-network processing.*" They proposed to carry out the processing task on the way, i.e., somewhere in between the devices and the Cloud. Advancement of network devices in terms of processing and storage capacity has fueled the concept to be effective. These devices are generally placed at the edge of the network, hence also called edge device. Due to close proximity to the IoT devices, real-time processing can be achieved. And since most of the data are being processed before being sent to the Cloud, network congestion has reduced significantly. In-network data processing would be exceptionally rewarding for the sensor networks.

Since the sensor nodes expense most of their energy in sending and/or receiving data streams [7], transferring all the sensed data to a remote site would squeeze the battery power of the devices [8]. As the concept has been popularized, people have come up with different "in-network processing" strategies, although the fundamental objectives of all of them are same. The main reason to coin different terms to describe the same concept is the lack of standard agreement on network edge [9]. In the following sections, we shall discuss such three architectures named *Fog computing, Mobile-edge computing,* and *Cloudlets.*

3.4.3 Fog Computing

Cisco has emerged as the largest player in offering "in-network processing" solution that provides an environment for IoT data processing and analysis close to the devices [10, 11]. Cisco has given a fascinating term of this computing paradigm – Fog Computing – as if the cloud has come down to the earth (edge of the network) as fog. The processing is done on the edge devices such as switches, routers, set-top boxes, etc., which are placed at the edge of the local area network. That's why this computing paradigm is also known as *Edge Computing.* Within the network, IoT data are collected, processed, and stored at Fog/edge node, also referred as the IoT gateway through which the Fog interacts with the Cloud. Fog computing is not meant to replace the Cloud rather complement it. The transient data that are required to be processed immediately for an instant response or the data that need to be stored for a short period only are processed in the Fog. If the IoT data need to be stored for a longer duration for further processing such as analysis and mining, they are forwarded to a Cloud server. Depending on the data type and processing type, different strategies are adopted by Fog computing [12]. Fog computing relieves the network from being overloaded by performing the preliminary processing at the ground, filtering out the unnecessary data, and letting pass only the data that are required to be further processed or stored. This approach has a far-reaching impact when large numbers of connected devices (according to Cisco that could be 50 billion by 2020) are linked to the Cloud. Interconnecting several context-related Fog nodes gives a panoramic opportunity for sophisticated and context-aware data analysis. Fog computing offers better privacy and reliability than Cloud. Connecting to the Cloud depends on the external network which may go down occasionally. Local computing facility will be a great prospect for mission-critical IoT applications in those unfortunate conditions.

3.4.4 Mobile-edge Computing

Mobile phones, especially smartphones, are becoming an integral part of IoT, both as the source and sink of data and services. Today's smartphones are boasted with a number of sensors which make smartphones as a prominent outlet of data generation, along with social network data. From the end-user perspective, events are notified to the smartphones as well as users may instruct and control the "things" through their smartphones. If the IoT devices could be connected to the mobile networks directly (i.e., avoiding going to the Cloud), the mobile users' quality of experience (QoE) would be significantly improved. Considering that, lately, MEC (Mobile-Edge Computing[1]) [13], a concept fostered by ETSI (European Telecommunications Standards Institute), has been considered as the potential platform for IoT data processing. MEC is a distributed computing environment where computational exertions are shunted to the mobile-edge rather than the core network. Here, the mobile-edge refers to a position, such as an LTE macro base station or a 3G Radio Network Controller (RNC) site, that is considered as the edge of the network [14]. In MEC, the base stations provide computing and storage facilities, in contrast to traditional base stations which just forward the traffic but do not perform any computational activity [15]. To carry out computational works, MEC employs a computing facility, known as MEC Server, at (or near) the base station. Besides providing connectivity, the MEC Servers provide computing and storage resources also. Instead of at the Cloud, IoT applications and services are deployed at the MEC Server that is much closer. Mobile users' requests land on the MEC Server and processed over there, rather than being forwarded to the Cloud. Similarly, IoT data are also processed and analyzed at the nearest MEC Server and the results are directly served at the users' mobile phones. This would reduce processing and communication latency, and consequently the response time to a great extent and that should certainly enhance the user's QoE. It is true that the resources at the base

---

[1]Recently, ETSI has decided to change the name "Mobile Edge Computing" to "Multi-Access Edge Computing," keeping the same acronym, as they are endeavoring to comprise Wi-Fi and fixed access technologies as well to their specifications [31].

stations are not comparable to that of offered by a Cloud data center in terms of capacity, scalability, and reliability. But accessing services deployed at the Cloud would put an immense burden on the mobile network considering that the scale at the smartphones is used in the IoT and the bandwidth limitation in a user network. MEC pacifies this load off the core mobile network by placing computing resources and services and caching and/or compressing content at the edge of the network. MEC Servers are equipped with filters and rule sets, so they can act as filters by consuming most of the raw data at the edge, thus saving bandwidth substantially [15]. Since the network edges are sited nearby to the mobile subscribers, MEC pledges to offer an efficient and quality service. MEC Servers provide access to user traffic and real-time RAN (radio and network) information that can be tapped to dole out personalized and context-aware services to the mobile users. The local information congregated at the base station allows MEC to provide location-aware services to the users by associating them with local points-of-interest, businesses, and events [14].

3.4.5 Cloudlets

The concept of Cloudlet has emerged from the convergence of mobile computing and Cloud computing. Developed at Carnegie Mellon University, Cloudlet is envisioned as the middle-tier between mobile devices and Cloud [16]. It aims to slash end-to-end latency by getting Cloud computing services closer to the edge. A Cloudlet is a miniature Cloud data center which emulates the Cloud facility much closer to the devices. A Cloudlet is ideally, one hop (Wi-Fi) distant from the mobile devices. It allows just-in-time response to computing intensive mobile applications. Architecturally, Cloudlets possess adequate computing power in the form of multicore CPU cluster, RAM, and cache and offer Cloud-like virtualization to run the computational jobs from mobile devices. They carry out the resource-intensive computational jobs delegated by resource-constrained mobile devices. A client application installed on the mobile device searches for the nearest available Cloudlet. The executable jobs are offloaded to the Cloudlet, cached, processed, or aggregated there and the result is sent back to the device or forwarded to the Cloud. Cloudlets are generally well connected to a Cloud server. They do not store the computations and results but only cache temporarily (hence referred as a soft state) and transfer them to the Cloud if necessary. But what happens if no Cloudlet is found in the proximity? Does it fail? No, not exactly. The application will react in such a way that there were no Cloudlets ever.

Either the job will be executed on the device only or it will be routed to the Cloud. In the latter case, the application has to tolerate the expected delay. As mentioned in the previous section, smart mobile devices, equipped with a number of sensors, have made IoT ubiquitous and have become a significant source of IoT data. Hence, Cloudlets are becoming a genuine potential for processing these huge near-real-time data, having said that the potential of Cloudlets need not be limited to mobile IoT applications only. Cloudlets can be leveraged to process data from other IoT devices as well, provided having suitable interface and middleware.

3.4.6 On-site Processing

On-site processing is a technique that is typically deployed on the sensor devices so that the sensed data are processed in and by the sensor devices themselves. This contrasts with the other approaches mentioned earlier, where the source of the data and the processing unit are separate. Typically, the sensor devices do not have enough computing resource; that is why the sensed data are driven to some external system for processing. In the previous sections, we have seen few options that we have when it comes to real-time or near-real-time data processing provided that there is available required infrastructure in the proximity. But ubiquitous and pervasive consumption of IoT may get us into situations where neither such computing facility is available nearby nor there is enough bandwidth to send the data across to a Cloud server. Let us consider two separate cases [15]:

1. Commercial jets generate 10 TB for every half an hour of flight which should be analyzed for various flight-related operations including to run the flight in auto-pilot mode.
2. Offshore oil rigs generate 500 GB of data weekly and required to be analyzed to monitor pipelines and seismic sensors.

In the first case, data processing should be in hard real time, and generally, there is no in-network processing facility available within the aircraft. In the second case, data may be processed in soft real time but in such environments, the available bandwidth for data transmission is very limited. So, sending a huge amount of data to a remote location for processing is exceedingly strenuous. In both the environments, it is necessary for the sensor devices to have *in situ* data processing capability. This will save a lot of energy and will lengthen the lifetime/operating period of the devices. To execute the tasks like data aggregation, the sensor devices may solicit the help of their close neighbors. Also, if most of the data are processed locally, they do not have to be deported to the upper layers. The volume of data transported over the network will be considerably lessened, which becomes decisive where communication facility is not enough. On-site data processing should also be a preferable choice in those IoT applications where IoT data need not be archived, centrally processed, or coalesced from multiple sources.

3.4.7 In-memory Computing

In-memory Computing (IMC) is quite a different computing archetype in essence than of those discussed prior. The concept of IMC is not new at all but it was not explored extensively until recently. In traditional computing, data are brought to the computational unit which is a time-consuming and resource-intensive process [17]. The traditional way of data processing is to first store the data on the hard disk and then transfer a part of it to the primary memory (RAM) from where processor would fetch it (probably via cache) for processing. After completion, the processed data are sent back to the hard disk and the subsequent part is loaded into the RAM and the cycle continues until the entire chunk of the data on the disk that is to be processed is completed. The major bottleneck in this process is the data transferring delay (a) from disk to RAM and vice versa and (b) from CPU to RAM and vice versa. So, it is obvious that that total processing time will increase commensurate with the data in the disk to be processed. That is why disk-based technologies are seemed to be impotent in sustaining the exigencies of real-time IoT data deluge. IMC tries to curb the latency by transposing the traditional computing principle. It aims to bring the processing unit where the data are in the RAM, usually bypassing the disk. It is realized by placing the processor and RAM very close to each other, favorably on the same chip. By this tactic, data movement will be minimized which results in reduced memory latency. Also, the lack of data movement keeps the system bus free and always available. All these gains ensure a significantly bettered processing speed. This is the reason why IMC has been considered as the most appropriate solution for real-time IoT data processing. IMC should be ideal, especially, to generate live events out of streaming data.

To analyze the streaming data, we require on-the-fly processing systems those can process real-time data without storing them on the disk. In fact, IMC is probably the only means to deal with the in-flight data efficiently [17]. DRAMs (dynamic RAM) are sufficiently faster than data stream sources to handle the data in the flow. If a sufficient number of DRAMs are employed and utilized pertinently, astonishing speed performance can be achieved. The larger the RAM size, the faster the processing. A cluster of these high-speed DRAMs will be capable of providing us with a super fast high-performance computing amenity. With memory cost dipping 30% per year [18], the performance–cost ratio of DRAM has dramatically increased. This has aided organizations to attain lower TCO (total cost of ownership) for data processing systems with a consummate performance experience. The utilization of high-performance in-memory computing is not merely limited to streaming data processing but it can be leveraged to apply real-time analytics to operational datasets as well [18]. At the rate the organizations are flooded with data from different sources including IoT, much of the insights may remain unearthed if the processing capability is not at par. IMC enables businesses and organizations to discover patterns and solutions in near real time and in timely decision making. By identifying and analyzing customer preferences and behaviors in near real time, businesses can dish upwell-timed service providing improved QoS, thus enriching customers' QoE. Adopting IMC, the possibilities organizations can draw from its data are endless and can attain business benefits that were infeasible before [17]. Most of the IT giants like Oracle, Microsoft, SAP, IBM, Tableau Software, Tibco Software, etc., are offering Big Data solutions, especially Big Data analytics tools based on IMC platform [19]. Organizations can build in-house in-memory data computing with the help of Amazon Spark that offers open source platform for in-memory data processing. GridGain provides an in-memory processing platform that works for both data grid and computing grid [20].

We assume that by now few questions might be knocking on your judicious mind. First is very obvious that DRAM being a volatile memory how does it pledge to not losing data? The simple resolution is to keep backups and sync them continually. An alternative option is to use non-volatile RAMs. Second, can't we apply the same logic to have in-cache computing? After all, the cache is at least three times faster than DRAM, so we could have even faster processing. Theoretically, that would be a wonderful option but the physical limitation of cache is the main obstructive factor why in-cache computing is not feasible, at least for the current computer architecture [20]. And since that is not going to change in foreseeable future, IMC will continue to be a great prospect for real-time data processing and analysis in coming years.

3.5 Security and Privacy Issues Involved in Each Archetype

Grid computing: Since in grid computing, the IoT data remain within the organization and behind the firewall, there is a minimal concern for security and privacy. Intrusion detection techniques can be applied to any probable external threat. Also, there is the least possibility of being affected by a DoS (denial-of-service) attack. But if the grid is made of inter-organizational and inter-administrative domains, the security and privacy may be compromised.

Cloud computing: Cloud is a public service; hence, the security and privacy concerns are always there. There is every possibility of a security breach for on-transmission data. Privacy is not guaranteed since data are located on a foreign server. One noteworthy problem in Cloud known as account hijacking is that fraudulent people access the services that are billed to other's account by hijacking one's account [21]. This can be really damaging to organizations as it can hamper their integrity and reputation. The organization may incur a significant loss if confidential data are leaked or forged. In another form of attack called service traffic hijacking, an attacker eavesdrops on the users' activities and transactions if they get access to users' credentials. They can manipulate or divert the processable data to undesirable locations. Another more common form of assault that the Cloud has to deal with more often is the DoS attack where Cloud services are denied to the legitimate users. Hence, processing IoT data will not be a good idea with respect of security unless strong security measures are adopted at the Cloud end.

Fog computing: In Fog, the preliminary processing is done within the network. Hence, it is supposed to be more secure than Cloud. When the pre-processed data are sent to the Cloud for storing and further processing and analyzing, it is exposed to the threat. But Fog devices which are at the edge of the home network and connect to the public network are vulnerable. Attackers may target these devices to inject malicious code into the system. They can introduce man-in-the-middle attack by replacing (virtually) genuine Fog device by the forged ones [22]. So, it is essential to protect these gateway devices in the Fog.

Mobile-edge computing: MEC servers are not really in-house properties. If these servers are forfeited, data security is also be conceded. Also, in MEC, mobile networks are used for data communication which is not hard to intercept. That is a concern for data privacy.

Cloudlet: The security and privacy aspects of Cloudlets are very much similar to Fog. Unless the Cloudlets and the gateway devices in the Cloudlets are not breached, the IoT data should be secure. But like Fog, the data might be unprotected when they are transported to the central Cloud.

On-site processing: On-site processing is assumed to be the safest among all others if standard devices are used and unless the hackers are able to penetrate really deep into the system.

In-memory computing: Memory servers are generally in-house properties of organization (otherwise, it will dilute the advantage of In-memory computing). So, it may be considered as a safe option for IoT data processing. But as usual, if the processed output is sent to the central repository, it might be exposed to a security threat.

3.6 A General Discussion on the Security and Privacy Issues in IoT

IoT has attracted ample attention of people from every sector. Everyone is in the flurry to taste the water and be among the forerunners in wringing the juice, without much planning and consideration. But it will be hazardous to overlook the additional security menaces that it brings to the cyber world. In the consumer IT space, IoT has immense promise in the areas of Smart Grid, intelligent transportation system, smart healthcare, intelligent water and waste management, and intelligent public safety and surveillance. All these above components form an integral part of the smart city that many governments are now planning to launch. Consequently, due to the inherent communication ability of these smart devices, the consideration of security and privacy of the exchange of data between the device and the controlling unit becomes paramount. As healthcare industry is likely to be a major beneficiary of IoT-enabled devices, the protection of patient sensitive data is critical for the wider adoption of IoT.

Most of the IoT devices are easy to access that makes them vulnerable to attacks. Security attacks like data tampering, deactivation, and tag detaching can make the scenario really challenging for IoT systems along with usual networking threats such as spoofing, eavesdropping, denial of service, etc. The key vulnerable areas are – an insecure web interface, insufficient authentication/authorization, insecure network services, and lack of transport encryption. IoT exposes users to identity theft and security breaches. IoT devices are easier to hack compared to the other typical target devices (e.g., computers, tablets, and smartphones), mainly because IoT devices are with a limited capacity and cannot afford to run protective applications (e.g., complex encryption and decryption techniques and anti-virus software) on their own. Hackers require minimal resources to bring down any IoT system, say, by launching DDoS (distributed denial of service) attacks. IoT botnets are becoming as favorite armament to carry out this and are shoving their feet even deeper with the proliferation of IoT adoptions.

The existing security technologies may be handy to some extent but they are not enough. For example, since the IoT devices shall communicate with each other and with the controller using the telecom network or other open networks, the current network authentication mechanisms (e.g., IMEI [23] and ESN [24]) may not be sufficient. Novel approaches and technologies (e.g., blockchain [25]) are required to ensure the privacy and security of IoT data. Because of the heterogeneity of the devices, it is essential that we are prepared for new attacks and plan innovative defenses. Government and organizations should deploy risk assessment frameworks (e.g., NIST [26] and COSO [27]) and opt for an effective policy framework for safeguarding their sensitive data. Countries are at different stages of implementation of regulatory guidelines depending upon their threat perception. The regulators play a crucial role in operationalizing IoT framework including the guidelines for data transfer and storage. Governments must step into the scenario for regulating and controlling IoT security. The Government not only should set security standards but also set standards for the IoT devices that come into the market and frame strict policies and rules that are to be followed by all the stakeholders [28]. Though this will be difficult to implement for a globally networked system like Internet due to several in intercontinental obligations, technologies such as blockchain can play a big role in it thanks to its qualities that include completely decentralized and P2P, autonomous, open, and secure [29].

3.7 An Off-the-wall Outlook on Security and Privacy Concerns for IoT

As more people are coming across IoT, more we are hearing about the dangers of its security vulnerability. The popular presumption is that the widespread and pervasive adoption of IoT with the lack of security standards, measures, and practices offers the picnic party to the hackers. They can explode the refrigerator at home, take control of our car while we are driving, set the whole city into the dark by taking hold of the Smart Grid, get hold of weapons of our soldiers, and so on [30]. In short, they can take down anything that has been connected. But aren't we reacting aberrantly? Since inventing the Internet and the WWW, we have been connected. Have we really faced that kind of cyber devastation? Then why are we losing nerve now? Not every IoT systems are that risky and some of them require the least attention in that aspect. The security risk should be assessed based on the IoT application. For example, suppose someone is playing remotely with my home electricity connection by connecting and disconnecting my home line to the Grid and in another case, he has put the whole city into darkness for a week by making an irreparable fault in the Grid. Of course, both cases are scary but the second case is of more concern. Consider another case that it would be much more concerning if somebody takes control of my car while I am driving rather than if someone has hacked into the car's music system and playing it a whole night while it is in the garage. The point is that it is to be understood that every IoT use case has different security apprehension, and instead of being frenzied about the whole IoT, appropriate precautionary measures should be applied case basis based on the degree of the risk.

While we are more concern about the fictional ghoul who can get hold of the IoT and we shall be doomed, we may have been overlooking the bigger menace pertaining to our privacy. IoT has set its foot in every territory of our life including our bedroom. It records our eating and sleeping habits, it records our health status and medical history, it records our purchasing behaviors, it records where we are traveling, and it records our digital foot-prints and what not! In short, it keeps an eye on whatever we are doing in our daily life. If unscrupulous people can get their hands on these data, we would be perturbingly exposed to them. Using advanced analytical tools, they can uncover every detail of us of which we ourselves may not be aware of. And to do this, it is not necessary to be always illegal. Because the data are everywhere, the hunter just has to tap in the right place. So, instead of that imaginary devil who can breach the security of our IoT devices, we should be more concern about all those existing devils that are after our personal data [30].

3.8 Conclusion

IoT has influenced our lives in a great way and in future also it will continue to do so. Future IoT devices will be truly ubiquitous and will be used more pervasively than today. The increasing number of IoT devices will generate abundant of data. To utilize IoT to its true capacity, these data are to be processed effectively. For that, suitable data capturing and processing architectures are needed. In this chapter, we have discussed a few of such architectures. We have seen that processing IoT data is a challenging job. We have also assessed the security and privacy aspects of each architecture. We have seen if the data are processed within the home network (e.g., Grid and Fog), the risk is less because the longer the data stays at the public network (i.e., Internet), the more it invites the risk. If the data are sent to the remote centralized server (Cloud) through the outer network, the risk is maximum. We also reasoned that not only the IoT security, we should also an emphasis on conserving data privacy. There should be standardized policy and framework for IoT applications in organizational as well as Government levels. As a concluding remark, for a good reason, it will be wise to refrain from connecting everything to the IoT that can be.

References

[1] Zhou, Z., Liu, M., Zhang, F., Bai, L., and Shen, W., "A Data Processing Framework for IoT based Online Monitoring System," in *IEEE 17th International Conference on Computer Supported Cooperative Work in Design*, 2013.

[2] Ma, M., Wang, P., and Chu, C. H., "Data Management for Internet of Things: Challenges, Approaches and Opportunities," in *IEEE International Conference on Green Computing and Communications and IEEE Internet of Things and IEEE Cyber*, 2013.

[3] Tracey, D., and Sreenan, C., "A Holistic Architecture for the Internet of Things, Sensing Services and Big Data," in *13th IEEE/ACM International Symposium on Cluster, Cloud, and Grid Computing*, 2013.

[4] Kovatsch, M., Lanter, M., and Duquennoy, S., "Actinium: A RESTful Runtime Container for Scriptable Internet of Things Applications," in *3rd International Conference on the Internet of Things (IOT)*, Wuxi, China, 2012.

[5] Wang, Q., Lee, B., Murray, N., and Qiao, Y., "CS-Man: Computation service management for IoT in-network processing," in *27th Irish Signals and Systems Conference (ISSC)*, June 2016.

[6] Kovatsch, M., Mayer, S., and Ostermaier, B., "Moving application logic from the firmware to the cloud: Towards the thin server architecture for the internet of things," in *Sixth International Conference on Innovative*

Mobile and Internet Services in Ubiquitous Computing (IMIS), Palermo, Italian, 2012.

[7] V Cantoni, V., Lombardi, L., and Lombardi, P., "Challenges for Data Mining in Distributed Sensor Networks," in *18th International Conference on Pattern Recognition (ICPR)*, HongKong, 2006.

[8] Gaber, M. M., "Data Stream Processing in Sensor Networks," in *Learning from Data Streams: Processing Techniques in Sensor Networks*, 41–48, 2007.

[9] Orsini, G., Bade, D., & Lamersdorf, W., "Computing at the Mobile Edge: Designing Elastic Android Applications for Computation Offloading," in *8th IFIP Wireless and Mobile Networking Conference*, 2015.

[10] Bonomi, F., Milito, R., Zhu, J., and Addepalli, S., "Fog Computing and Its Role in the Internet of Things," in *MCC'12*, Helsinki, Finland, 2012.

[11] Cisco White Paper, "*Cisco Fog Computing Solutions: Unleash the Power of the Internet of Things*," Cisco, 2015.

[12] Cisco White Paper, "*Fog Computing and the Internet of Things: Extend the Cloud to Where the Things Are*," Cisco, 2015.

[13] Hu, Y. C., Patel, M., Sabella, D., Sprecher, N., and Young, V., "Mobile Edge Computing: A key technology towards 5G," *ETSI (European Telecommunications Standards Institute)*, 2015.

[14] Huawei, IBM, Intel, N. Networks, N. DOCOMO and Vodafone, "*Mobile-Edge Computing – Introductory Technical White Paper*," 2014.

[15] Li, P., "*Semantic Reasoning on the Edge of Internet of Things*," Master's Thesis, University of Oulu, Oulu, 2016.

[16] Satyanarayanan, M., Simoens, P., Xiao, Y., Pillai, P., Chen, Z., Ha, K., and Amos, B., "Edge Analytics in the Internet of Things," *Pervasive Computing*, 24–31, 2015.

[17] Industry Perspectives, "*Why In-Memory Computing Technology Will Change How We View Computing*," available at: http://www.datacenter knowledge.com/archives/2014/01/06/memory-computing-technology-will-change-view-computing/. [Accessed 27 December 2016].

[18] GridGain White Paper, "*The GridGain In-Memory Data Grid*," GridGain Systems, Inc., 2016.

[19] Sultan, "*Top 10 In-Memory Business Intelligence Analytics Tools*," 2015, available at: www.mytechlogy.com/IT-blogs/9507/top-10-in-memory-business-intelligence-analytics-tools/#.WGIWWlN97Dc. [Accessed 2 January 2017].

[20] Ivanov, N., Stamper, J., and Sterin, I., *"In-Memory Computing: Driving the Internet of Things,"* available at: https://www.gridgain.com/resources/webinars/memory-computing-driving-internet-things [Accessed 27 December 2016].

[21] Lord, N., *"What is Cloud Account Hijacking?,"* available at: https://digitalguardian.com/blog/what-cloud-account-hijacking [Accessed 10 April 2017].

[22] Wang, Y., Uehara, T., and Sasaki, R., "Fog Computing: Issues and Challenges in Security and Forensics," in *IEEE 39th Annual International Computers, Software and Applications Conference*, 2015.

[23] Saha, A., and Sanyal, S., "Survey of Strong Authentication Approaches for Mobile Proximity and Remote Wallet Applications - Challenges and Evolution," *International Journal of Computer Applications,* 108, 2014.

[24] Turgut, B., and Caglayan, M. U., "An AAA based solution for secure interoperability of 3G and 802.11 networks," in *New Trends In Computer Networks*, Imperial College Press, 368–383, 2005.

[25] The Economist, *"Blockchains: The great chain of being sure about things,"* 2015, available at: https://www.economist.com/news/briefing/21677228-technology-behind-bitcoin-lets-people-who-do-not-know-or-trust-each-other-build-dependable [Accessed 17 January 2018].

[26] "National Institute of Standards and Technology," *United States Department of Commerce,* available at: https://www.nist.gov/ [Accessed 17 January 2018].

[27] "Welcome to COSO," *Committee of Sponsoring Organizations of the Treadway Commission*, available at: https://www.coso.org/Pages/default.aspx [Accessed 17 January 2018].

[28] Dickson, B., *"How insecurity is damaging the IoT industry,"* available at: https://bdtechtalks.com/2016/10/23/how-insecurity-is-damaging-the-iot-industry/ [Accessed 8 April 2017].

[29] Banafa, A., *"How to Secure the Internet of Things (IoT) with Blockchain,"* available at: https://datafloq.com/read/securing-internet-of-things-iot-with-blockchain/2228 [Accessed 17 January 2017].

[30] Linthicum, D., *"IoT Security: It's More About Privacy Than Killer Roombas,"* available at: https://www.rtinsights.com/iot-security-its-more-about-privacy-than-homicidal-roombas/ [Accessed 20 December 2016].

[31] Morris, I., *"ETSI Drops 'Mobile' From MEC,"* available at: http://www.lightreading.com/mobile/mec-(mobile-edge-computing)/etsi-drops-mobile-from-mec/d/d-id/726273 [Accessed 27 December 2016].

4

Safeguarding the Connected Future: Security in Internet of Things (IoT)

Priti Maheshwary and Timothy Malche

Department of CSE, Rabindranath Tagore University (formerly known as AISECT University), Bhopal, MP, India

Overcoming security issues are one of the main challenges for the Internet of things (IoT). There are different security issues that are related to IoT in the context of sensor nodes, system network, IoT cloud, IoT backend database, IoT applications, system software, and firmware. IoT applications have real-time requirements and expected to deliver a high level of reliability. The IoT applications operate in safety–critical environment and therefore need justification extreme security measures. This chapter discusses the need for security in IoT, IoT architecture and attacks on each layer of the IoT architecture, attack surface in IoT, IoT protocols, built-in security features in protocols, and security management. The aim of this chapter is to provide a thorough understanding of major security concerns that are necessary in building secure IoT.

4.1 Introduction

Internet of things (IoT) is a technological revolution that serves as a global network of intelligent physical objects. In this network, all intelligent objects exchange information and services over the Internet where some information or services are private and some are publicly accessible. In IoT, these objects are called "things." A "thing" can be any object embedded with sensors, electronics, software, and network connectivity. A "thing" has an IP address and has the ability to transfer data over the Internet. Some examples of IoT applications include connected human body, connected home, connected industry, connected environment, connected city, etc. At present, there are billions of objects exist on Earth. Through IoT, all kinds of objects ranging from simplest (such as coffee cup, umbrella, cloths, etc.) to more complex (such as watch, car, refrigerator, washing machines, airplanes, smartphones, computers, etc.) will be able to connect, interact, and communicate. IoT is a very complex platform for connecting things. There are many objects, sensors, actuators, and technologies available for building the IoT ecosystem. It is estimated that there will be around 50 billion connected objects by the year 2020. At present, 80% of the connected objects are exposed to security threats and vulnerability. In the IoT ecosystem, all objects are connected to the Internet all the time sensing and sharing information. In order to be more useful, IoT devices must make real-time bi-directional connections to the Internet instead of only uploading the data. Therefore, this type of communication must be secure enough; otherwise, it will be easier for the attacker to get into the system and gain control over it and steal sensitive information. Demand for security also raises when IoT is used in many important domains such as medical, military, and even many smart city applications, where data are more sensitive and systems are more open for remote access and control. Thus, it becomes very important to understand security requirements when designing the IoT system in order to make our connected world safe and secure.

This chapter discusses the IoT architecture and security needs. However, the main focus of this chapter is to understand security issues in the IoT which will help in developing more secure connected future. This chapter is divided into the following sections:

- Need for security – This section discusses about security challenges, confidentiality, integrity, availability, and non-repudiation.
- IoT architecture – This section discusses about the fundamental IoT architecture such as Device–device, Device–Cloud, Device–Gateway, Gateway–Cloud, and Cloud–Backend–Applications.
- IoT protocols – This section discusses about IoT protocol stack layers such as link layer, network layer, transport layer, application layer, and protocols at each layer.
- Security classification and access control – This section discusses about the classification of data used in IoT and also privacy issues in IoT.
- Attack surface in IoT – This section discusses about the attack surface in IoT from hardware and software to communication links between objects.

- Security features in IoT protocols – This section discusses built-in security features of protocols on transport layer such as SSL/TLS and DTLS and protocols on application layer such as MQTT, CoAP, XMPP, and AMQP.
- Security management – This section discusses two of the most important security management techniques, identity and access management and key management.
- IoT-based smart home and security issues – This section discusses IoT smart home and security issues in the smart home.

4.2 Need for Security

IoT security is the area of major concern. Security in IoT deals with safe-guarding connected objects and networks in the IoT. The IoT involves the growing collection of objects known as things with unique identifiers. These objects have the ability to automatically transfer data over a network. Most of the increase in IoT communication comes from computing devices and embedded sensor systems which are used in industrial machine-to-machine (M2M) communication, home and building automation, vehicle-to-vehicle communication, smart energy grids, and wearable devices. The main problem is that because the idea of connecting all objects to the Internet is relatively new, and therefore security has not always been measured in product design. Most of the IoT products sold have old and unpatched embedded operating systems and software and users often fail to change the default passwords on smart devices or fail to select sufficiently strong passwords. To improve security, an IoT device that is directly accessible over the Internet should be distributed in the network and have network access restricted. The distributed components should also be monitored to identify potential risk, and actions should be taken if problem arises. According to Proofpoint [1], an enterprise security firm, the 25% of connected devices are IoT botnet. The botnet was made up of devices other than computers, including baby monitors, smart TV, and household appliances, etc., which require more security.

4.2.1 Security Challenges in IoT

4.2.1.1 Vulnerability points

As more and more devices are getting connected to the IoT, these devices can create a potential security and privacy risk as they are the potential doorway for the IT infrastructure and data. A study by Fortify shows that 70% of the devices connected in the IoT today are vulnerable to the security issues.

4.2.1.2 Privacy

There are many devices that are connected in the IoT and can collect some valuable and personal information of the user or surrounding environment from the sensor device or from the application. Some of the devices even transmit this information via the network or the Internet without any encryption or proper authentication. This makes the data extremely vulnerable to misuse and theft if it is accessed by an unauthorized person. Privacy is the major concern in the IoT system like Smart Home, Smart Grid, Smart City, Defense System, etc. Many information in the Smart Industry and Research must also require privacy of data.

4.2.1.3 Authentication and authorization

The passwords require of sufficient complexity and length. Weak and easy passwords add to the vulnerability. A large number of users configure their devices and accounts with weak passwords and they do not even change the default passwords given by vendors for the devices. All IoT devices and application which allow users to control objects in the IoT infrastructure must also require proper authentication and authorization to make sure that the system is monitored and controlled by the right person.

4.2.1.4 Transport encryption

In IoT, the devices are connected and communicate with each other by transferring data from one device to another device over the Internet. Many of the devices do not encrypt the data and hence an addition to the security issues. Transport encryption is where information which is being sent from one device to another device is in an encrypted form. Transport encryption is very crucial because most of the devices are transmitting data all the time. Many devices failed to encrypt data, even when the devices were using the Internet. Encryption is needed for the given amount of information which is being shared between the device, the cloud, and mobile applications.

4.2.1.5 User interface

Web interfaces of the devices can also raise security concern. Some issues in the web interface that can raise problem are persistent cross-site scripting, poor session management, weak default credentials, etc. Using these methods, hackers are able to identify valid user accounts and take control over them using password reset features.

4.2.1.6 Firmware update

IoT devices do not use encryption when downloading software or firmware updates. IoT devices also require proper authorization for updating the firmware of devices. If the firmware or software is not protected, an attacker can easily gain control over the device and then the whole network as the device, in this case, will open a door for the attacker to enter in the system and access to confidential information [2].

4.2.2 Confidentiality, Integrity, and Availability

Confidentiality, integrity, and availability also known as CIA triad is a model which is designed to guide policies for information security. This model is also sometimes referred to as the AIC triad (availability, integrity, and confidentiality) to avoid confusion with the Central Intelligence Agency. The three elements of the triad are considered the most crucial components of security.

Here the confidentiality is a set of rules that limit access to information, integrity is the assurance that the information is trustworthy and accurate, and availability is a guarantee of reliable access to the information by authorized people. Three of these also play an important role in IoT since the information and data in IoT is also crucial [3].

4.2.2.1 Confidentiality

Confidentiality is used to prevent sensitive information from reaching the wrong people. In order to make sure that the right people can reach it, the access must be restricted to authorized users to view and access the data. The data must be categorized and stringent measures should be implemented according to those categories.

The IoT-related confidentiality issues are unauthorized access, use, disclosure, and modification of private and sensitive information. The eavesdropping attacks in the IoT infrastructure give rise to confidentiality issues. Eavesdropping, at the communication level, refers to intentionally listening to private conversations over the communication links. This attack may provide sensitive information to attacker from the unencrypted or unprotected communication. The information that transmits between either two IoT devices or between the IoT client and the IoT server may contain information such as username, password, access control information, device configuration information, shared network password, and other sensitive information which is critical to the system. The attacker tries to steal information that transmits between two IoT devices or IoT devices and cloud to gain control over the IoT system. In this attack, the attacker takes advantage of unsecured network communications to get access to the data being sent and received in the IoT infrastructure. It is difficult to detect eavesdropping attacks because these attacks do not cause network transmissions to appear to be operating abnormally. Therefore, to avoid such an attack, the communication must be encrypted and protected.

4.2.2.2 Integrity

Integrity includes maintaining the consistency, accuracy, and trustworthiness of data during its entire life cycle. Data must not be changed in transit, and necessary steps must be taken to ensure that data cannot be altered by unauthorized users. These measures include file permissions and user access controls. Version control can also be used to prevent erroneous changes or accidental deletion by authorized users. Some policies also require detecting any changes in data that might occur as a result of non-human-caused events such as an electromagnetic pulse (EMP) or server crash. It is also secure to use cryptographic checksums on data for verification of integrity.

The military application of IoT may be taken as an example for integrity. The Military IoT sometimes also known as M-IoT is the application of IoT where all objects of the battlefield such as troops, tanks, weapons, vehicles, aircrafts, etc., are equipped with various sensors and connected to each other via the Internet and share real-time data. In the military IoT, the military gathers data from various sensors on a variety of platforms such as aircraft, weapon systems, ground vehicles, and troops in the battlefield. The sources of data in military IoT are command, control, intelligence, surveillance, and reconnaissance systems that process and disseminate the mission-critical information, such as the position of an incoming threat. IoT-connected sensors and radars collect and transmit data of the position and movements of troops and enemies and other events in the battlefield. The IoT for military makes it attractive and more useful but it also makes such a framework vulnerable. In M-IoT, it becomes necessary to maintain a secure environment which supports safe exchange of information with accuracy, trustworthiness, and in real time. Therefore, all the information shared in the battlefield must be maintained for data integrity.

As illustrated in Figure 4.1, the location information of the battlefield by the sensor on an aircraft is sent to soldiers in real time. This information

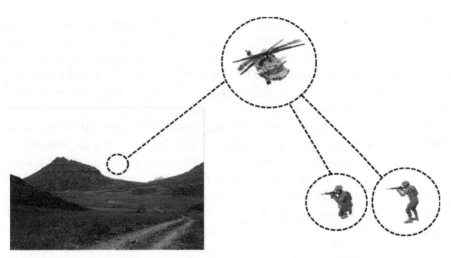

Figure 4.1   Location-based communication at battlefield.

has to be precise and thus such communication requires integrity. Although M-IoT requires various measures of IoT security, the following are in relation to integrity:

Table 4.1   Integrity measure in M-IoT location-based communication application

Security Measure	Action
Consistency	Data delivery in the battlefield.
Accuracy	Location data, alerts, and commands from control unit
Trustworthiness	Origin of data from sensors and control unit

The software integrity of IoT devices as discussed by Echard [4] may also be considered in the military application of IoT as it focuses on detection and prevention of modification of the original software of the IoT devices, so that the device cannot be used for unintended purposes because these devices are used to send vital information of the battlefield upon which further necessary actions may be taken.

4.2.2.3 Availability

Availability can be ensured by carefully maintaining all required hardware, performing hardware repairs straight away whenever it is needed, and maintaining appropriately working operating system environment which is free of software conflicts. Availability is very much important in IoT because all IoT devices that are equipped with sensors require be up and running all the time. It is also important to keep the system devices updated with the current version of software of the firmware. An adequate communication bandwidth is also required for IoT devices to transmit data and preventing the occurrence of bottleneck. Redundancy, failover, RAID, and even high-availability clusters can mitigate serious consequences when hardware issues occur; therefore, fast and adaptive disaster recovery is essential for the worst case scenarios. Safeguards against data loss or interruptions in connections should also be included in case of unpredictable events such as natural disasters and fire. Extra security equipment or software such as firewalls and proxy servers can also be used to guard against downtime and unreachable data due to malicious actions such as denial-of-service (DoS) attacks and network intrusions.

4.2.2.4 Non-repudiation

A non-repudiation service provides an assurance of the origin of data or delivery of data so that it can protect the sender against false denial by the recipient that the data have been received, or to protect the recipient against false denial by the sender that the data have been sent. Non-repudiation is the ability to prove that an event has taken place, so that this cannot be repudiated later. For example, in the e-mail system, non-repudiation is used to guarantee that the recipient cannot deny receiving the message and that the sender cannot deny sending it. Non-repudiation is a method that guarantees message transmission between objects via digital signature, encryption, etc. It is one of the five pillars of information assurance (IA). Non-repudiation is also an important aspect in the IoT. Because the data in IoT are most important, it becomes more crucial in real-time systems such as disaster and recovery system, industry and research, healthcare, etc., where the data alone play an important role in the system. In such scenario, non-repudiation service plays an important role.

Digital signature, a cryptographic function, can be used to implement non-repudiation services in an IoT system. Digital signature also provides integrity, authentication, and data origin protections. It is designed to be unique to the signer, the individual or device responsible for signing the message and who possesses the signing key. There are two types of digital signatures, one which uses symmetric key (secret, shared key) and other which uses asymmetric key (private key is unshared). In Figure 4.2, an IoT device that sends the message adds its signature to the message. The message being send is now called signed message and any other IoT device with the appropriate key can perform the inverse of signature operation

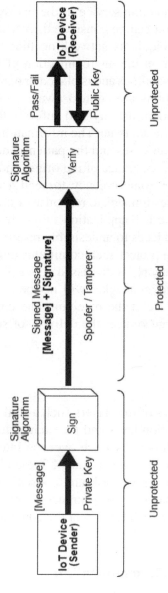

Figure 4.2 Digital signature protection from spoofing and tampering.

called signature verification. If the signature verification process fails, then the verifier should not trust the integrity of data and if they have originated from the trusted source [5].

To understand non-repudiation service in the IoT system, an example of healthcare sensor application can be considered. A healthcare sensor application provides wireless ubiquitous networking functionalities. Healthcare sensor application is based on the interconnection of tiny nodes that have sensing and/or actuating capabilities and is placed or planted in a human body. Healthcare sensor applications are context-aware, personal, and dynamic in nature. It senses the patient activity and informs the doctor on remote location about the patient's health issue. The information may be vital if some emergency medical condition arises and the patient requires attention. Thus, such an application requires assurance of the origin and delivery of data, for example, if a doctor senses some urgency to send a decision on patient's health condition that requires immediate response or action either by patient himself or by actuators. The IoT application failing in such scenario would affect doctor's decision and leads to undesirable results that may cause harm to the patient. Therefore, to protect such communication, a digital signature may be used. A digital signature, as discussed above, is helpful in protecting such communication so that an attacker will not be able to falsify data and since it is protected, a device at the receiving end cannot deny its origin. Therefore, such protection guarantees the delivery of such vital information about the patient.

4.3 IoT Architecture

The IoT architecture consists of four layers such as link layer, network layer, transport layer, and application layer and the architecture consists of sensor devices, gateway devices, a cloud service over the Internet, communication devices and medium, etc. As shown in Figure 4.3, each layer of the IoT architecture is responsible for a specific task and consists of different protocols. The IoT architecture also has various communication methods. This section discusses the IoT architecture on the basis of the communication model in the architecture.

4.3.1 Device-to-Device Communication

This model represents two or more devices that directly connect and communicate to each other, rather than through an intermediary application server. These devices communicate over many types of networks, including IP

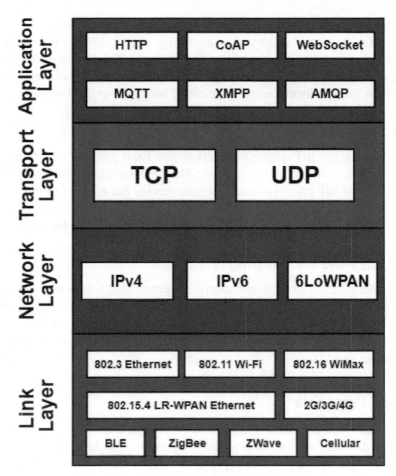

Figure 4.3 IoT layers and protocols.

networks or the Internet. The example of device-to-device communication is a smart light bulb controlled by smartphone soft buttons. The other example may be vehicle-to-vehicle communication in smart transportation systems where one car sending warning message to other car for the short distance between the two cars.

4.3.2 Device-to-Cloud Communication

In this model, the IoT device directly connects to the Cloud service for IoT. The cloud service for IoT is responsible for collecting and storing data from sensor devices, setting rules, and providing processing on those data.

The cloud service may provide many services for the IoT devices such as sharing data over social network, generating alert, etc. This method takes advantage of existing communication mechanisms such as wired Ethernet or Wi-Fi connections to establish a connection between the device and the IP network which finally connects to the cloud service over the Internet.

4.3.3 Device-to-Gateway

In this model, the IoT device connects to the local gateway device and the gateway device connect to the Internet. This method adds extra strength and security to the IoT infrastructure. The gate device can be used to filter information by having a firewall. The IoT gateway device acts as an intermediary between the IoT sensor device and the cloud service over the Internet.

4.3.4 Back-end Data-sharing

This model refers to a communication architecture that allows users to export and analyze data collected from an IoT sensor device and uploaded to cloud to the other services or other data sources. This architecture provides a way for granting access to the uploaded sensor data to third parties, organization, services, or users [6].

4.4 IoT Protocols

The TCP/IP protocol stack is at the heart of the Internet. The protocol stack in IoT is divided into four layers. These layers are link layer, network layer, transport layer, and application layer. Each layer is responsible to perform a specific task with the help of different protocols. Figure 4.3 represents the four layers and protocols used at each layer.

4.4.1 Link Layer

This layer is responsible for transferring data between IoT devices. This layer hides the details of underlying hardware and represents itself to the upper layer as the medium to communicate. The link layer converts data stream to signals bit by bit and send it to the underlying hardware. At the receiving end, this layer receives data from the hardware in the form of electrical signals, assembles it in a recognizable frame format, and hands over to the upper layer. The link layer works between two hosts which are directly connected.

This direct connection could be point to point or broadcast. The link layer performs the following tasks:

Framing: The link layer takes packets from the network layer and encapsulates them into frames. Then, it sends each frame bit by bit on the hardware. At the receiving end, the data link layer picks up signals from hardware and assembles them into frames.

Addressing: The data link layer provides hardware addressing mechanism. Hardware address is assumed to be unique on the link.

Synchronization: As soon as data frames are sent on the link, it synchronizes the transfer for both ends.

Error Control: Whenever signals encounter problem in transition, it detects errors and attempted to recover actual data bits and also provides error reporting mechanism to the sender.

Flow Control: IoT devices on the same link may have different speeds or capacities. The link layer ensures flow control that enables communicating devices to exchange data on the same speed.

Multi-Access: There may be a high probability of collision. The link layer provides mechanisms such as CSMA/CD to ensure that data are shared among multiple devices. The link layer consists of the following protocols which determine how the data are physically sent over the network. The following are the IEEE standards in the link layer.

802.3 Ethernet is a collection of wired Ethernet standard for Ethernet networks. It defines the media access control (MAC) of wired Ethernet. It has 802.3 as coaxial cable, 802.3.i as copper twisted-pair connections, 802.3.j as fiber optics connections, 802.3.ae as fiber, etc. These standards provide data rate from 10 Mb/s to 40 Gb/s and even higher. *802.11 Wi-Fi* is a set of specifications for wireless LAN (WLAN) technology. 802.11 specifies an over-the-air interface between a wireless client and a base station or between two wireless clients. 802.11 has several specifications that are used in IoT. The standard 802.11a operates at 5 GHz, 802.11b/g operates at 2.4 GHz, 802.11n operates at 2.4/5 GHz, 802.11ac operates at 5 GHz, and 802.11ad operates at 60 GHz. These provide data rates from 1 Mb/s to 6.75 Gb/s. *802.16 WiMax* is a series of wireless broadband standards which provide data rates from 1.5 Mb/s to 1 Gb/s, mobile stations provide till 100 Mb/s and fixed station till 1 Gb/s. *802.15.4 LR-WPAN* is a technical standard which defines the operation of low-rate wireless personal area networks (*LR-WPANs*). It provides data rates around 40 Kb/s to 250 Kb/s. This standard provides low-cost and low-speed communication suitable for powered constrained IoT devices. These are very useful for remote sensing data. *2G/3G/4G* are the mobile communication standards where 2G includes GSM and CDMA, 3G includes UMTS and CDMA2000, and 4G includes LTE. IoT devices based on these standards can communicate over a cellular network. These standards provide data rates from 9.6 Kb/s to 100 Mb/s [7].

This layer is highly prone to several attacks on IoT devices. These attacks are the following:

- *Address Resolution Protocol (ARP) spoofing* is a technique by which an attacker sends ARP messages onto the network with an aim to associate the attacker's MAC address with the another host such as gateway causing traffic of IP sent to the attacker. This attack allows the attacker to intercept data frames on network, modifies and stops the traffic, and opens door for other attacks.
- *Denial of Service (DoS) attack* is an attack in which an attacker makes sensor node or network resource unavailable by disturbing the service.

4.4.2 Network Layer

It is also called the Internet layer in the IoT which is responsible for sending IP datagrams from the source network to the destination network. This layer performs host addressing and packet routing. The datagram contains source and destination addresses which are used to rout them from the source to the destination across multiple networks. The host identification is done using hierarchical IP addresses scheme such as IPv4 or IPv6, etc. The network layer is responsible for performing the following tasks [8]:

- Addressing devices and networks.
- Populating routing tables or static routes.
- Queuing incoming and outgoing data and forwarding them according to quality of service constraints set for those packets of data.
- Internetworking between two different subnets.
- Delivering packets to destination with best efforts.
- Providing connection-oriented and connectionless mechanism.

The network layer consists of protocols which are considered useful in IoT such as IPv4 which is the most deployed internet protocol. This protocol uses 32-bit addressing schemes that allow the total 2^{32} addresses. But as more and more devices connected to the Internet, these addresses got exhausted in 2011. As a result, a new version of protocol comes into play. The IPv6 is the newest version of internet protocols and a successor of IPv4. This protocol uses 128-bit addressing schemes that allow the total 2^{128} addresses. Another protocol is *6LoWPAN* a somewhat contorted acronym that combines the latest version of the internet protocol (IPv6) and low-power wireless personal area networks (LoWPAN). This protocol enables smallest devices that have a limited processing ability to transmit information wirelessly using an internet protocol. It operates on a 2.4 GHz frequency range and provides data transfer rates around 250 Kb/s. Thus, this protocol is considered more useful than the other for resource- and power-constrained devices.

The network layer vulnerabilities in IoT are categorized as the following attacks:

- *Routing attack* in this attack an attacker may change the routing information to create routing loops which significantly deteriorates quality of service since the routing information is not encrypted.
- *Eavesdropping attack* is an attack when an attacker can gain access to a communication channel. It is a passive attack unless the attacker modifies the received packets and sends it back to the source. This method is called replay attack and it is a very common subtype of spoofing.
- *Replay attack* is an attack in which an attacker obtains a signed packet, and apart from whether he can decrypt it, it gains the trust of the destined entity by re-sending the packet at a later time.
- *Sinkhole attack* is an attack in which some nodes are made more attractive to network traffic than other normal nodes. When packets reach the sinkhole node, the messages may get dropped and forwarded with altered content.
- *Wormhole attack* is an attack in which a wormhole is maliciously made, low latency link, over which the attacker can replay messages. The attacker receives packets at one point in the network and sends it to another point in the network, and then replays it from that point.
- *Sybil attack* is an attack in which the attacker uses sensor nodes or devices with multiple identities. This results in traffic which seems to have many sources. This attack corrupts resource usage, redundancy, or voting concepts originally present in the infrastructure.
- *Node replication:* In this attack, the attacker copies the identity of a sensor node and creates another virtual sensor node with the same identity in order to send false through random routes to disrupt the network.

4.4.3 Transport Layer

The transport layer provides peer-to-peer and end-to-end connections between two processes on remote hosts. The transport layer takes data from the upper layer (i.e., application layer) and then divides it into smaller size of segments, then numbers each byte, and hands over to the lower layer (network layer) for delivery. This layer performs the following functions:

- It breaks the information data provided by the application layer into smaller units called segments. It numbers every byte in the segment and maintains their accounting.
- It ensures that data must be received in the same sequence as they were sent.
- It provides end-to-end delivery of data between hosts which do not belong to the same subnet.
- All server processes that communicate over the network are provided with Transport Service Access Points (TSAPs) also known as port numbers.

The transport layer has two of the most important protocols in reference to IoT. *TCP (Transmission Control Protocol)* is the connection-oriented protocol. TCP enables two devices to establish a connection and exchange streams of data. TCP guarantees delivery of data and also guarantees that packets will be delivered in the same order in which they were sent. It helps avoiding network congestion and congestion collapse. It provides reliable, ordered, and error-checked delivery of a stream of data between applications running on hosts and communicating by an IP network. *UDP (User Datagram Protocol)* is the connectionless protocol. It is used for time-sensitive applications. It is a transaction-oriented and stateless protocol which does not guarantee packet delivery. It provides fast data transmission with a constant dataflow but does not provide reliability.

This layer is also vulnerable to the following attacks related to IoT:

- *Session Hijacking* is also known as TCP session Hijacking. It is a way of taking control over user session by secretly obtaining user's session ID and pretending to be the authorized user for accessing the data.
- *TCP Land Attack* is the attack in which the attacker sends an SYN packet to the host server which usually has an open TCP port.
- *UDP Flooding Attack* is the attack in which the attacker floods the server machine with countless requests; this makes the machine to think that the attacker, which pretends to be the authorized user, really needs service urgently and the server machine starts providing the services to the attacker; as a result, the users who actually need the service are often overlooked.

- *TCP and UDP Port Scanning* is a technique in which the attacker performs the port scanning of various tools of the host machine to find the open ports on the machine and after the ports are identified, it is used to attack the server.

4.4.4 Application Layer

The application layer is the top layer where the actual communication is initiated and reflects. This layer does not serve any other layers but takes the help of transport and all layers below it to communicate or transfer its data to the remote host. This layer prepares communication between devices for data transmission over the network and initiates the data transfer. It receives data from the network and prepares it for use. The application layer consists of various protocols suitable for IoT devices. The protocols at this layer are *HTTP (Hyper Text Transfer Protocol)* which is the stateless protocol that follows the request–response model and uses URI (Universal Resource Identifiers). *Constrained Application Protocol (CoAP)* is a transfer protocol designed for use with constrained devices and constrained networks in the IoT. The protocol is designed for M2M applications such as smart energy and building automation. It runs over UDP instead of TCP and follows the request–response model. *WebSocket* is another computer communication protocol which provides full-duplex communication channels over a single TCP connection. It is based on TCP and it allows streams of messages to be sent back and forth between the client and the server.

Message Queuing Telemetry Transport (MQTT) is a lightweight messaging protocol which is very useful in resource-constrained environment. This protocol simply distributes telemetry information over the network. It uses a publish/subscribe communication pattern, and is most widely used for M2M communication. It plays an important role in the IoT. *XMPP (Extensible Messaging and Presence Protocol)* is a protocol which is based on Extensible Markup Language (XML). This protocol is used mostly for instant messaging (IM) and online presence detection application. It is also sometimes called jabber protocol. It is used for real-time communication and streaming of XML data. It is most widely used in applications like gaming, multi-party chat, and voice/video calls. It is a decentralized protocol which is very useful for IoT devices. *Advanced Message Queuing Protocol (AMQP)* is an open standard message-oriented protocol. It supports both point-to-point and publisher subscriber models and is used for routing and queuing data over the network.

The application layer in IoT is also open for several attacks as the following:

- *Malicious code injection:* The attacker injects malicious code from unknown location into the system and tries to steal or manipulate the data of the authorized user.
- *Denial-of-service attack:* The attacker pretends to be an authenticated user who logs into the system and interrupts the working of the network.
- *Phishing attack:* The attacker gains credentials of the system and accesses the system and damages the data.
- *Sniffing attack:* The attacker uses a sniffer application to attack the network of the system and read the unencrypted packets.

4.5 Security Classification and Access Control

The following are the security classification and access control that must be the subject of consideration in the IoT [9].

4.5.1 Data Classification

In information security, data classification depends upon its level of sensitivity. It is also depending on the level of impact such that data should be disclosed, altered, or destroyed without authorization. The main security controls which are appropriate for safeguarding data are determined with the help of classification of data. All data should be classified into one of the three sensitivity levels, or classifications such as *Restricted Data* are said to be restricted when the unauthorized disclosure, alteration, or destruction of that data could cause a considerable level of risk. An example of restricted data is data for defense organization. The highest level of security controls should be applied to restricted data. *Private Data* can be classified as private when the unauthorized disclosure, alteration, or destruction of that data could result in a moderate level of risk. By default, all data that are not explicitly classified as restricted or public data may be treated as private data. A reasonable level of security controls must be applied to private data. *Public Data* can be classified as public when the unauthorized disclosure, alteration, or destruction of that data may result in little or no risk. Examples of public data are press releases, course information and research publications, weather information, rail/flight information, etc. There may be little or no controls required to protect the confidentiality of public data. The level of control is required to prevent unauthorized modification or destruction. The levels of controls are authentication, authorization, and data integrity. *Authentication* is the way of protecting the data. Authentication is a process in which the credentials provided by users are compared with those available in the database of authorized users. Authentication is important in IoT applications such as Smart Home. Only the authenticated user may be able to access the Smart Home system. *Authorization* is the process of providing access rights to data and resources related to information security. It is also used to specify access control in a particular IoT system. Authorization is required for many IoT applications in order to specify who has the access to the system. The example is the Smart Grid system which requires only the authorized user to access the system. *Data integrity* is the maintenance and the assurance of the accuracy and consistency of data. Data integrity applies to the entire life cycle of data. It is a critical aspect to the design, implementation, and usage of any IoT system which stores, processes, or retrieves data. Because data are the most important in IoT systems upon which actions are taken and decisions are made, the data integrity plays an important role in this context.

Authentication and data integrity can be achieved by using digital signature as described in Section 2.2.4. The authorization may also be achieved using such digital signature as person with valid signature will only be granted to have an access right on a particular IoT device in the IoT infrastructure.

MQTT protocol can also be used to achieve authorization. MQTT, publish/subscribe protocol, allows clients to write and read topics. In this process, all MQTT clients do not have permission to write the topics; similarly, all do not have permission to read. A control can be put using an MQTT broker that restricts such permission to MQTT clients by keeping access control list to topics. The MQTT broker can authorize MQTT clients using client ID or any other technique and grant access to topic by topic lookup to determine whether MQTT clients are authorized to read, write, or subscribe to topics.

4.6 Attack Surface in IoT

The attack surface in IoT is a doorway for attackers to enter into the IoT system and steal information and even take control over the system. Thus, for making IoT systems secure, these surfaces must be identified, protected, and restricted from malicious and unauthorized access [10].

4.6.1 Interfaces

Embedded devices have a number of circuits which enable to exchange information. This information is exchanged using the agreement on common protocols that can be divided into two groups, serial and parallel. Serial and parallel are the interfaces that allow multiple bits of data exchange and require a large number of pins, whereas in a serial interface, one bit of data is transferred at a time, which is slower than parallel, but has an advantage of requiring a less number of pins. Some of the popular examples of protocols which use serial communication are Universal Serial Bus (USB), Ethernet, UART, SPI, and I2C. Serial interfaces in any IoT device are fairly useful for the attacker because it provides a doorway for the attacker to step in the system and interact with the device even if the device appears totally locked. There are different interfaces available that may vary from device to device.

The debugging test pins on the interface of the IoT device are most effective physical hardware attack vectors available to an attacker. The most known interfaces are *UART* (Universal Asynchronous Receiver/Transmitter) which is popularly known as a serial interfacing protocol which allows data to be transferred without the requirement of external clock and thus it is called asynchronous. UART acts as an intermediary between a parallel and a serial interface. It is particularly useful to sniff debug and boot up messages and logs, obtain shell access, etc. In the real-world IoT devices, UART grants to directly obtain unauthenticated root access which in turn allows an attacker to enter into the system. *SPI* (serial peripheral interface) is, unlike UART, a synchronous serial communication interface. It requires a clock to facilitate communication. It can be used for dumping the firmware from the device. SPI works on the master–slave architecture. In this architecture, one device sends data and the other device receives data. The pinouts on SPI are named MISO (master out slave in) and MOSI (master out slave in). *I2C* (inter-integrated circuit) also works on the master–slave architecture and allows multiple slave chips to communicate with one master. The usage is pretty similar to SPI [11]. *JTAG* is a hardware access point that enables one or more devices to connect over externally facing pins for debugging purposes. JTAG is used for two main purposes, first is called boundary scan and second is called debug access. The boundary scan is a simple form of debugging that makes sure that all of the devices are connected correctly. Debug access provides a way to talk to intelligent devices and catch its raw live computation. This is very powerful since it allows the access to registers, memory contents, interrupts, and the ability to pause and redirect flow of instruction. When an IoT device has some secret on it, this technique can use debug access to actually read the contents of memory and pull that secret out of the memory chip. Some CPUs have JTAG interfaces built into it and some don't. Many board designers expose these CPU capabilities with easy to access pins for debugging purposes [12].

4.6.2 Software and Cloud Components

The software and cloud components are also available for attackers as the entry point for the IoT system. Therefore, in order to secure the system from an attack, it must be well understood. The *Firmware* of the device is an important tool for the attacker to enter into the IoT system. An attacker may be able to find valuable information, user ID and password of the network, in the firmware using hex editor or other tools. The insecure update of firmware can also lead to security issues. Since the firmware is present on the IoT sensor device, the attacker can damage the device and thus prevent any data from being submitted to the server or facilitate wrong data to be submitted to the server, which can lead to no information or error prone information displayed to user. This must be protected using cryptography.

Web Application Dashboard is also an important area of concern. The dashboard of the IoT application is the control unit of the overall IoT system. If somehow an attacker gains control to the dashboard, much harm can be made to the whole system. This kind of attack must be protected with the strong password encryption technique and secure session management. *Mobile Interface* provided by most IoT applications to its users to give users an easy way to access the system using their smartphones is also an entry point to the attacker if it is not protected with some strong security techniques.

4.6.3 Radio Communication

Since the IoT sensor devices mostly rely on radio communication such as Wi-Fi, BLE, Cellular, Zigbee, ZWave, and 6LoWPAN. The insecure radio communication also allows an attacker to gain access to the system. Many attacks such as denial of service (DoS), password-based attacks, wireless phishing, man-in-the-middle attack, compromised-key attack, sniffer attack, wireless intruders, rogue APs, endpoint attacks, and evil twin APs can be carried against the radio communication. Thus, it must be protected using cryptography and firewalls.

4.7 Security Features

There are various protocols used in IoT at each of the discussed layers (Figure 4.3). Each layer performs a specific task using these protocols. There are some security features which are already built in to some of these protocols which are discussed in this section.

4.7.1 TLS/SSL

Transport layer security (TLS) and its predecessor, secure sockets' layer (SSL), are collectively known as SSL. These are the cryptographic protocols that offer communication security over a computer network. Many versions of the protocols are used in applications such as web browsing, email application, Internet faxing, instant messaging, and voice-over-IP (VoIP). Websites use TLS to secure all communications between servers and web browsers. The transport layer security protocol provides data integrity and privacy between two computer applications that communicate with other. The communication of client–server which is secured by TLS has one or more of the properties such as connection which is private or secure because symmetric cryptography is used to encrypt the transmitted data. The server and client agree upon the details of the encryption algorithm and cryptographic keys that are used before the first byte of data is transmitted. This negotiation of a shared secret is secure and reliable. The identity of the client–server that communicates with each other is authenticated using public-key cryptography. The connection ensures integrity because each message transmitted undergoes message integrity check using a message authentication code to prevent undetected loss or modification of the data during transmission [13]. There is a need to relate to the IoT case. How is the TLS used in IoT? Which server (cloud?) and which client (e.g., which device)? Credit card info on the web is using TLS, then is there anything IoT that requires TLS?

TLS has an important role in IoT. The MQTT protocol, discussed in next section, is often implemented to operate over TLS. The MQTT broker can be configured for certificate-based authentication of the MQTT client. Using it, the MQTT broker maps the information in the MQTT client X.509 certificate to determine the topics to which the client has permission to subscribe or publish. The real-world example is IBM Watson IoT platform's MQTT API that encrypted communications on ports 8883 or 443. The platform requires TLS. The registration of devices on the platform requires the use of the TLS connection, as the MQTT password is transmitted back to the client which is protected by the TLS tunnel [5].

4.7.2 MQTT (Message Queuing Telemetry Transport)

MQTT protocol is a publish/subscribe messaging protocol designed for M2M communication often known as IoT protocol. MQTT protocol consists of an MQTT client, which can be any microcontroller device that has an MQTT library running on it and an MQTT broker to which the client connects over the network. The MQTT client can be a publisher or subscriber or both. The MQTT broker is the heart of publish/subscribe activity and can handle up to thousands of concurrently connected MQTT clients and is primarily responsible for receiving messages, filtering them, and then sending the message to all subscribed clients. The MQTT broker provides one or more topics that allow an MQTT client to publish/ subscribe a message to a related topic. The topic is generally a small token that may store valuable information. MQTT security has three fundamental concepts, identity, authentication, and authorization. Identity is about providing identification for being authorized and given authority. Authentication is about proving the identity and authorization is about managing the identity. Identity – Identify an MQTT client by its client identifier, user ID, or public digital certificate. One or other of these attributes defines the client identity. The MQTT server authenticates the certificate sent by the client using the SSL protocol, with a password set by the client. Using this server can control the resources which the client can access based on the identity of the client. The MQTT server identifies itself to the client with its IP address and digital certificate. The MQTT client uses the SSL protocol to validate the certificate sent by the MQTT server.

Authentication is done by both the MQTT server and client. A client authenticates a server with the SSL protocol. An MQTT server authenticates a client with the SSL protocol, or with a password, or both. If the client authenticates the server, but the server does not authenticate the client, then the client is known as an anonymous client. It may be possible to establish an anonymous client connection over SSL, and after that, it authenticates the client with a password which is encrypted by the SSL session. The client can be authenticated with a password rather than a client certificate, because of the certificate distribution and management problem. The MQTT client server communication is safe and secure as it happens in the following steps.

- The MQTT client authenticates the server to make sure that it is connected to the right server. It does it by the server certificate with the SSL protocol.
- The MQTT server verifies that it is connected to the correct client. It does by authenticating the client certificate with the SSL protocol, or by authenticating the client identity with a password.

Authorization is not part of the MQTT protocol but it is provided by MQTT servers. Authorization depends on the server in what the server does. MQTT servers are publish/subscribe brokers. The MQTT authorization rules on the server control the clients that can connect to the server and topics which a client can be publish or subscribe [14].

4.7.3 DTLS (Datagram Transport Layer Security)

Datagram transport layer security (DTLS) is a communication protocol that provides security for datagram-based applications. For this purpose, it allows the applications to communicate to prevent eavesdropping, tampering, or message forgery. The DTLS protocol is based on the stream-oriented transport layer security (TLS) protocol. The DTLS provide a similar security guarantee to TLS. The DTLS protocol datagram preserves the semantics of the transport so that the application has not to suffer from the delays associated with stream protocols, but it has to take care of packet reordering, loss of datagram, and data larger than the size of a datagram network packet. There are three main elements when considering security, namely, integrity, authentication, and confidentiality. DTLS can achieve all of them. DTLS solves two problems, reordering and packet lost. DTLS adds three implements: first, packet retransmission, second, assigning sequence number within the handshake, and third, replay detection. Unlike network layer security protocols, DTLS in the application layer protects end-to-end communication. The end-to-end communication protection will make it hard for attackers to access to all text data that pass through a compromised node. DTLS also avoids cryptographic overhead problems that occur in lower layer security protocols.

4.7.4 CoAP (Constrained Application Protocol)

Constrained Application Protocol (CoAP) is now becoming the standard and popular protocol for IoT applications. Security is always a major concern to protect the communication between devices. It is a lightweight protocol designed for M2M communications within IoT applications. CoAP uses Datagram Transport Layer Security (DTLS) as communication protocol. CoAP is by default bound to UDP and optionally to DTLS, providing a high level of communication security [15].

4.7.5 XMPP (Extensible Messaging and Presence Protocol)

Security has always been a key issue for XMPP. The security considerations in XMPP are high security, certificate validation, client–server communication, server–server communication, protocol layers, firewalls, and base64. *High security* is the mutual confidentiality which is maintained in both sides of communication when a certification-based authentication is provided. If the certificate is issued, then only the authorized certificate should be accepted. *Certificate validation* is when a certificate is issued, it should be reviewed by both the communicating objects. In *client–server communication*, the client must support the TLS and Simple Authentication and Security Layer (SASL) protocols to connect to the server. The encryption of the XML stream is done using TLs and the authentication is supported out by the SASL. After the service is verified, then only the client should communicate to the server. The IP address which the client communicates should be kept private so that the channel is prevented from intruders.

In *server–server communication* unlike the client, the server should support the TLS and SASL protocol for ease of communication. Authentication and integrity of the data are ensured with the use of SASL in the server communication. *Protocol layers* consist of four protocols in order to be used in XMPP as mentioned in the security implementation. The protocols are TCP, TLS, SASL, and XMPP, where TCP is the base layer where the connection between the client and the server is established. TLS encrypts the XML stream, SASL provides authentication, and XMPP is the application layer. For *Firewalls*, TCP is widely used for XMPP communication. For client-to-server communication, port 5222 is used and for server-to-server communication, port 5269 is used. *Base64* helps to recognize the trusted client and server. The server validates the client and if any data are found irrelevant, then they are not accepted. This improves the data integrity and passing the correct data.

4.8 Security Management

It is the set of functions that protect the data against unauthorized access, control the activity of user operations, set up security rules, monitor security events, etc. There are many security management techniques used of which two of the most important are discussed here which can be useful for IoT.

4.8.1 Identity and Access Management (IAM)

It is the security discipline that enables the access of a right resource to a right individual at a right time for a right reason. This management is mission-critical and requires to make sure that proper access to resources across increasingly heterogeneous technology environments is granted and also to meet increasingly rigorous compliance requirements. It covers many issues related to security such as how users can achieve an identity, the protection of user identity, and also the technologies which support that protection (e.g., network protocols, digital certificates, passwords, etc.). IAM has the following functions:-

- The pure identity function which includes creation, management, and deletion of user identities without concern to access or entitlements.
- The user access (log-on) function which defines how proper users can gain access to the system with provided credentials.
- The service function which includes a system that delivers the user and devices personalized, role-based, online, on-demand, multimedia content, and presence-based services.
- Identity federation which includes a system that relies on federated identity in order to authenticate a user without knowing his or her password. Identity federation is made up of one or more systems that associate user access and allow users to log in based on authenticating against one of the systems acting in federation.

In addition to creation, deletion, and modification of user identity data, identity management also controls additional entity data for use by applications, such as contact information or location. The system capabilities of IAM are authentication, authorization, roles, delegation, and interchange. The *authentication* is a verification of users using a password or a biometrics device. *Authorization* is managing authorization information that defines who can access what operations in the context of a specific application. *Roles* are the groups of operations that users are granted. Roles are often related to a particular job or job function. For example, a user administrator is authorized to reset a user's password, while a system administrator may have the ability to assign a user to a specific server. *Delegation* allows local administrators to perform system modifications without a global administrator. It may also mean that one user can perform operation on behalf of other user. *Interchange* is the use of Security Assertion Markup Language (SAML) protocol to exchange identity information between two identity domains [16].

The IAM can be illustrated through the IoT-based smart parking system. The smart parking system is an IoT reference system because it contains multiple endpoints that capture data and send it to the database storage on the IoT server. The system also provides data analytics and decision making. In this example, the smart parking system, as discussed by Russell et al. [5], has the following features:

- Consumer-facing service: This service allows customers to locate the vacant parking spot and pricing.
- Payment flexibility: This service allows customers to pay for parking space using multiple payment methods such as credit cards, cash/coins, and mobile payment services.

- Entitlement enforcement: This service allows the ability to track time for purchased parking spot, determine the expiry time, sense the overstayed time at the parking spot, and communicate the violation to parking enforcement.
- Trend analysis: These services allow collecting and analyzing historical parking data and provide trend reports to parking managers.
- Demand-response pricing: This service allows the ability to change pricing depending on the demand for each space.

The security goals for the smart parking system could be:

- Maintaining the authentication of the customers and parking managers.
- Maintaining the authorization of who can change the parking spot pricing.
- Maintaining the roles of parking managers, attendants, administrators, and enforcement officers.
- Maintaining the delegation of administrators.
- Maintaining integrity of all data collected by the system.
- Maintaining confidentiality of sensitive data of the system.
- Maintaining the availability of the whole system and all of its components.

The functioning of the parking system is described in the following points:

A. The customer purchases the parking spot

- The customer installs the parking application on his/her smartphone.
- The customer registers the payment information to perform the transaction.
- The application provides real-time information of available nearby parking spots and pricing details.
- The customer selects the spots and drives to the spot.
- The customer uses the application to pay for the spot.

B. The parking enforcement officer is alerted to non-payment incidents

- The parking application records the parking session start time and also records the overstayed time spent by the vehicle at the parking spot.
- IP video cameras capture the video of the vehicle overstayed in the parking spot.
- The parking application correlates the video of the vehicle in the parking spot with start time and duration for parking transaction flags for video confirmation if the transaction duration has expired.

- The parking application then transmits an alert to enforcement application of the overstayed vehicle.
- The enforcement officer receives an SMS alert and proceeds to ticket the vehicle.

The following diagram shows the overall system architecture:

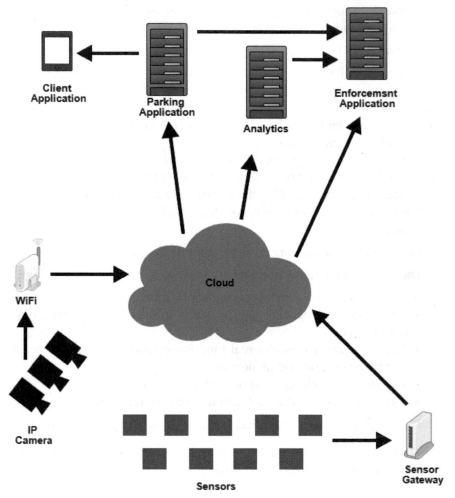

Figure 4.4   Smart parking system architecture [5].

The security controls that are needed to be implemented in the smart parking system are the following:

Table 4.2   Smart parking system security requirement

Attack on Smart Parking System	Type of Attack	Requirement of Security Control
The attacker gains control over customer's account and charges the customer for parking time	Spoofing	Authentication
Through unauthorized access, the attacker uses the parking spot for free of charge	Tempering	Authentication and integrity
The attacker uses the parking spot for free by claiming that the system has malfunctioned	Repudiation	Non-repudiation and integrity
The attacker gains access to customer's payment details	Information disclosure	Authentication and confidentiality
The attacker shuts down the smart parking system through the DoS attack	Denial of service	Availability
The attacker disrupts smart parking operations by implanting the rootkit on backend servers	Elevation of privilege	Authorization

4.8.2 Key Management

It is the technique used to manage cryptographic keys in a cryptosystem. It facilitates in generation, exchange, storage, use, and replacement of keys. KM includes cryptographic protocol designs, key servers, user procedures, and other relevant protocols. A key or cryptography key is a piece of information also called a parameter which determines the functional output of a cryptographic algorithm. Cryptographic systems may use different types of keys and one system may use more than one keys. The keys may be symmetric or private keys or asymmetric or public keys. In a symmetric key algorithm, the keys involved are the same for both encrypting and decrypting a message. Keys must be chosen carefully and also distributed and stored in a secured manner. Asymmetric keys, in contrast, are two distinct keys which are mathematically linked. They are typically used in conjunction to communicate. A Key Management System Security Policy provides the rules that can be used to protect keys and metadata. The key management is very important in IoT as it protects confidentiality, integrity, and availability and the process is secure [17].

4.9 IoT-based Smart Home and Security Issues

This section discusses Smart Home as the IoT model. A Smart Home can be defined an environment that has lighting, appliances, heating, air conditioning, TVs, computers, entertainment systems, and security and camera systems each of which is capable of communicating with one another. For example, a system to alert the user on his smartphone through a camera feed whenever a doorbell or letterbox is used or any unauthorized access to smart lock. All the connected "things" in a smart home can also be controlled remotely anytime, from any room in the home itself and as well as remotely from any location in the world by a phone or any device connected to the Internet. Because in smart home, all things are connected to each other via the Internet; it therefore also exposes security threats. The common security threats in smart home are confidentiality, authentication, and access. *Confidentiality* threat may result in unwanted release of private and sensitive information. *Authentication* threat can lead to tempering of sensing and control information. For example, through an unauthenticated alert, an attacker can confuse the house controller forcing him to open door into thinking of an emergency situation. *Access* is the greatest type of threat. It enables attackers to gain control of a system particularly at administrative level thus making system insecure.

A smart home may be vulnerable due to insecure firmware upgrade to IoT devices, unencrypted information sharing between the system components, etc. There are three of the most important security concerns for the smart homes. They must be carefully considered when designing the smart home to protect against being misused. First, smart home should not be equipped with *hackable connected devices*. As it has been observed, many of the devices in smart home used today are hackable. These devices include thermostats, smart TV, security camera, smart switch, smart door locks, home appliances, etc. All of these devices must be protected against all the possible attacks. If an attacker gains control over these devices, it can produce inconvenience to the user of the home. Even the computers and smartphones used to control the smart home system must be protected with a strong security policy. Second, smart home should not contain *compromised security systems*. The security system of smart home must also be protected using cryptography techniques, strong passwords, and firewalls.

Compromised security systems lead to exposing the security flaws in the smart home system. Imagine that if an attacker can gain control over the security camera on the door and record the door lock password entered by the user, then the attacker easily uses this information to access the home when the user is away from the home. Therefore, all the security systems in the house must be well protected. Third, smart home should not allow *Spying on Communication Systems*. Most of the home communication systems have devices such as video conferencing devices, computers, printers, phones, etc. Users are able to communicate with other gadgets and other people outside of their homes using such devices. The attacker can steal valuable information using passive attacks, or attacks that are used to gain unauthorized access without actually changing any of the data or code. This may result in monitoring telephone conversations and email messages and keeping track of how people are interacting with devices. Therefore, all the communication systems must also be strongly considered for security flaws.

4.10 Conclusion

The security is the major concern for IoT. Since billions of devices are getting connected to the Internet, the security concern for these devices is also rising. In this chapter, attack points including devices, interfaces, and software and cloud components are discussed, and communication layer security protocols that secure communication among IoT devices (clients and server, M2M, etc.) such as SSL/TLS and DTLS, MQTT, CoAP, XMPP, and AMQP are discussed here. Security is an important aspect of IoT as it helps to protect valuable data and avoided data misuse. The security concerns, as discussed in this chapter, must be thoroughly identified and implemented in all the IoT systems at action currently as well as being developed in the future. By enforcing these security concerns, an IoT system can be developed which is safe and suitable for better living, and thus the connected future can be safeguarded.

References

[1] *Proofpoint Uncovers Internet of Things (IoT) Cyberattack*, available at: http://investors.proofpoint.com/releasedetail.cfm?releaseid=819799
[2] *Securing the Internet of Things: A Proposed Framework*, available at: http://www.cisco.com/c/en/us/about/security-center/secure-iot-proposed-framework.html
[3] *Confidentiality, Integrity, and Availability (CIA Triad)*, available at: http://whatis.techtarget.com/definition/Confidentiality-integrity-and-availability-CIA

[4] Echard, C., Ensuring Software Integrity in IoT Devices. *J. Inform. Technol. Softw. Eng.* 7, 1–3, 2017.

[5] Russell, B., Van Duren, D., *Practical Internet of Things Security*, PACKT Publishing, 2016.

[6] *Understanding IoT Protocols*, available at: https://solace.com/blog/use-cases/understanding-iot-protocols-matching-requirements-right-option

[7] *Tutorials Point, Data Communication and Communication Network*, available at: https://www.tutorialspoint.com/data_communication_computer_network/data_link_control_and_protocols.htm

[8] *Tutorials Point, Data Communication and Communication Network*, available at: https://www.tutorialspoint.com/data_communication_computer_network/network_layer_introduction.htm

[9] *Data Classification Guidelines and Procedures*, available at: http://doit.niu.edu/doit/Policies/data-classification.shtml

[10] *Hardware Hacking for IoT Devices - Offensive IoT Exploitation*, available at: http://resources.infosecinstitute.com/hardware-hacking-iot-devices-offensive-iot-exploitation/#gref

[11] *Getting Started with IoT Security – Mapping the Attack Surface*, available at: http://resources.infosecinstitute.com/getting-started-with-iot-security-mapping-the-attack-surface/#gref

[12] *Why are JTAG and UART Still Effective Attack Vectors for IoT Devices?*, available at: https://p16.praetorian.com/blog/why-are-jtag-and-uart-still-effective-attack-vectors-for-iot-devices

[13] *Transport Layer Security*, Wikipedia, available at: https://en.wikipedia.org/wiki/Transport_Layer_Security

[14] *MQTT Security*, available at: https://www.ibm.com/developerworks/community/blogs/c565c720-fe84-4f63-873f-607d87787327/entry/mqtt_security?lang=en

[15] *Constrained Application Protocol for Internet of Things*, available at: http://www.cse.wustl.edu/~jain/cse574-14/ftp/coap/

[16] *Identity Management*, Wikipedia, available at: https://en.wikipedia.org/wiki/Identity_management

[17] *Key Management*, Wikipedia, available at: https://en.wikipedia.org/wiki/Key_management

5

Role of Blockchain in IoT Security

Tasneem Jahan and Imran Ali Khan

Department of Computer Science Engineering, Bansal Institute of Research, Technology and Science, Madhya Pradesh, India

Internet of Things (IoT) is a computing technology that has evolved to improve every object of our daily life, by transforming them into smart devices, thus by creating the automation of routine tasks and activities. It aims for designing a digital world to boost the business momentum and for the growth of industries. With the expansion of IoT across the globe, the major concern associated with it is its resistance to attacks and security. IoT deals with the scenarios that involve complicated network connections and computer capabilities for sensors controllers, and several other technologies to provide real-time computing services. IoT has heterogeneous collection of devices that vary in architecture and creates a varied network size. Security is an important aspect that is needed to be implemented efficiently by using protocols and algorithmic schemes. The following content of this chapter will cover the working of Blockchain, along with its advantages and disadvantages. Examples of Blockchain in IoT applications, security aspects of Blockchain, and Blockchain in business are also discussed.

5.1 Introduction

Internet of Things (IoT) security deals with the protection and safety of the devices and networks connected to the IoT. There are many things which we use in our daily life. These things are to be connected to the Internet using a wired or wireless backbone. IoT communication gets its data from the connecting devices which are used in industries' smart energy grids, home and building automation, vehicle-to-vehicle communication, and wearable computing devices.

The main problem is that the IoT industry is not focussing on the security measures, not only companies but the consumers also not concerned about the security threats. Furthermore, end users often fail to change the default passwords on smart devices – or if they do change them, fail to select sufficiently strong passwords. This gap can be fulfilled by a robust automated system which enables end users and companies to overcome such a kind of threat.

The objective of designing a secure IoT model has many obstacles for even the simplest tasks such as data sensing, data communication, and data storage. To achieve this goal from the scratch, the entire IoT system is classified into two discrete levels as:

1. System view: It groups various elements of an IoT ecosystem together, e.g., things or objects, network services, gateways, and cloud services.
2. Business view: It comprises the set of services offered by IoT, e.g., the platform architecture, applications, connectivity, and business model.

5.2 Current Trends and Their Challenges

The IoT system in its early evolution was based on a client/server model which is a centralized architecture. The entire communication must pass through the cloud servers. These servers authenticate the devices and connect them after identification to offer huge storage capacities and processing power. This centralized system suffers from the lack of flexibility for network expansion and rising costs to scale the infrastructure. Moreover, the centralized clouds require high maintenance costs, networking equipment, and larger server farms.

The expansion of IoT devices is required to hold the increasing amount of communications, and thus raises the subsequent cost to handle these communications. The major drawback of the centralized architecture is the bottleneck condition that arises because of increased load of communication and the low fault tolerance due to single point of failure. This eventually crashes the entire network. There are less technical experts for the IoT framework, and the centralized IoT model poses the issue of cost regarding life cycle management and its maintenance.

Hence, to address and face these issues of the centralized IoT model, there is a need to develop and improve security technologies. The concept of Blockchain is a promising technology and could answer the shortcomings of the centralized approach.

5.2.1 Blockchain Technology

Blockchain was invented in 2008 by Satoshi Nakamoto for use in the crypto currency bit coin, as its public transaction ledger. Blockchain can be abstracted as a database to keep track of ever growing records of data. The data here refer to the transactions involved in the business processes. The key feature of Blockchain is its distributed nature, (Figure 5.1) which means that there is no single or master device to hold a chain of transactions. It rather keeps a copy of chain of data records on each participating node.

The fundamental concepts of Blockchain are those which already exist. It uses the techniques of public key cryptography, hashing, and digital signature. Public/private key pair is the basis of communication. Any node initiating a transaction signs by its private key. The identity of nodes is their public key, rather than their IP address. Distributed ledger along with the process of hashed keys on the blocks makes Blockchain strong models among its competitors.

A Blockchain consists of two types of elements:

1. Transactions are the actions created by the participants in the system.
2. Blocks record these transactions and make sure that they are in the correct sequence and have not been tampered with.

5.2.2 Functioning of Blockchain

When a node initiates a transaction, and wishes to add that transaction in the chain, the rest of the nodes in the network will validate it. The validity is

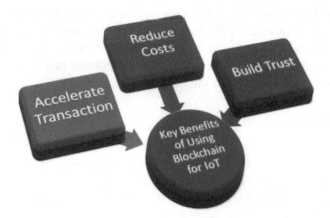

Figure 5.1 Blockchain in IoT.

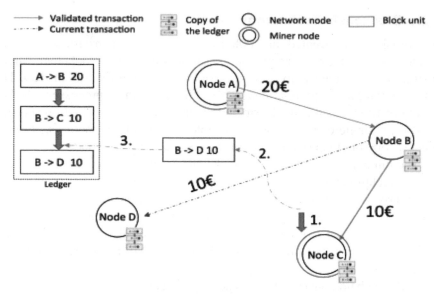

Figure 5.2 Transaction management in Blockchain.

verified by applying algorithms to the transactions. A transaction which is valid for one system may not be verified as valid by another system. Thus, the valid transactions differ among systems.

A block in Blockchain is a set of transactions which are valid, and then these transactions are sent to all the nodes in the network. They, in turn, validate the new block (Figure 5.2). Each next block that is being generated contains a unique fingerprint (hash) of the previously block.

In an IoT network, the Blockchain can keep an immutable record of the history of smart devices. This feature enables the autonomous functioning of smart devices without the need for centralized authority. As a result, the Blockchain opens the door to a series of IoT scenarios that were remarkably difficult, or even impossible to implement without it.

5.2.3 Construction of Blocks

Blocks are the basic elements of a database in a Blockchain. These are identified by the hash identifier. Each block contains a timestamp and list of its own transactions. A block can contain multiple transactions in the form of a Merkle tree. A Merkle hash is the root of Merkle tree.

Each node of the tree keeps a copy of the Blockchain and these copies are all in synchronization with each other, so as to ensure confidentiality and to confirm a benchmark that all nodes keep the same copy of transactions at the same level. This way, Blockchain manages all the security aspects of an IoT network by ensuring its integrity, transparency, authenticity, and availability.

By leveraging the Blockchain, IoT solutions can enable secure, trusted, and authentic messaging between devices in an IoT network. In this model, the Blockchain will treat message exchanges between devices similar to financial transactions in a Bitcoin network. To enable message exchanges, devices will leverage smart contracts which then model the agreement/disagreement between the two parties.

The "distributed ledger" concept of Blockchain is growing as a topic of great interest in the tech industry and beyond. The ledger assigns a valid tag to the transaction and arranges them according to the timestamp of their generation.

Blockchain technology offers a solution of recording transactions or any digital interaction in a way that is designed to be secure and transparent, and is also highly resistant to node failures, provides greater efficiency, and as such, it carries the possibility of renewing the business themes in industries and enabling new models for business processes.

Blockchains use complex mathematical functions to create a secure and definitive record of who owns what, when. In other words, Blockchains keep a ledger – which businesses can also use to track credits, debits, and other transactions.

Example: Cooperatively Owned Self-driving Cars

Using current technologies, a company like Uber (Figure 5.3) or Google maintains the servers necessary to run a self-driving car.

- Data are centralized with the service provider.
- Service providers compete to aggregate the most data to service the most customer.

Using a Blockchain-based service, any number of individuals could form an agreement between themselves to purchase a self-driving vehicle and share its maintenance among themselves (Figure5.4). Each cooperative group could form contracts with other groups and share usage of their vehicles among a wider group of peers.

Figure 5.3   Current system where we connect to a server for booking up a car.

- Individuals enter into smart contracts that define the ownership of machines, maintenance, requirement, and proper usage.
- Each group has their own agreement, and can enter into a new agreement with other groups.
- Reputation protocols allow for fluid exchange of information, services, and resources within and between groups.

These groups can set their own rules and enforce them using reputation standards. For example, a group could create a monthly maintenance checkup requirement that each must fulfill at least once per year – if they had not

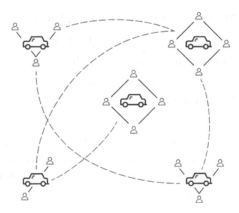

Figure 5.4   Blockchain system: Sharing machines.

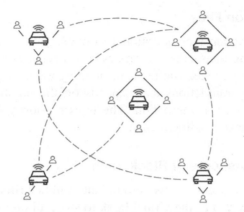

Figure 5.5   Blockchain system: Sharing machines like self-driving cars.

completed that duty, then the car would not unlock for that individual (Figure 5.5). The community could go a step further and block that individual's access to other services if that was encoded in their rules.

- Data are centralized with individuals and devices.
- Service providers compete to connect trusted peers.
- Marketplace improvement increases efficiency of sharing resources and determines fairer rate of exchange.

5.2.4 Bitcoin and Blockchain

The world of Internet uses electronic payment system and the money flows in the form of digital asset or crypto currency, known as a Bitcoin [1]. The system of Bitcoin is self-organized in the form of a transaction bill. The miners in the Blockchain collect only those transactions which are valid. These valid transactions are identified by a private/public key pair, designed using the hash algorithm.

The concept of BitCoin is mostly derived from the most promising feature of Blockchain, which is decentralization of verification of transactions along with suitable cryptographic procedures. The common peer-to-peer BitCoin system works on the shared Blockchain ledger. BitCoin uses digital signatures for maintaining the security and the Blockchain model facilitates the fully secured digital transactions.

5.2.5 Transaction Fees

Payment of transaction fees is optional to transactions. Miners handle the tasks of prioritizing transactions to process them according to those transactions that pay a higher fee. The fee is decided by the storage space required by the generating transactions and depends on the number of inputs that are required to create a transaction. The higher priority is given to those transactions that have pending inputs.

5.2.6 Security Aspects of Blockchain

The discussion in this text revolves around the "public" Blockchain, in which a ledger can be viewed as the virtual book to store all previous transactions. To handle the disorganization of continuously generating valid transaction blocks, Blockchain needs a protocol to achieve consistency of these blocks. There is a "miner" network node in the Blockchain network, whose role is to validate the new transactions. A new transaction is first validated by the miner before being added to the chain, and is then broadcasted to the entire IoT network. Blockchain applies a specific key to the block which contains valid transactions and then stores them into the distributed ledger. Validating a new transaction means verifying the entire ledger.

The miners are required to complete a proof of work (PoW) over all data blocks. This PoW is a hashcash function and it acts as a vote in the Blockchain to validate transaction log of Blockchain. This PoW makes it tougher for the nodes of the IoT network to predict which a computer generates the next block. Rather the Blockchain keeps on updating and every node can view the ledger.

5.3 Evaluating Blockchain

The present-day IoT solutions are expensive because of the high infrastructure and maintenance cost associated with centralized clouds, large server farms, and networking equipment. The communication cost is handled according to the pace of growth of IoT devices.

The issues identified in centralized approach are solved by the distributed ledger approach in decentralized mechanism of Blockchain [3]. It adopts a uniform peer-to-peer communication model to execute billions of transactions. It also offers the advantage of reduced costs of installing and maintaining data centers by distributing computation and storage needs across all devices of the IoT network, thus preventing the entire network to collapse even if a single node goes down.

Establishing a peer-to-peer communication model also presents a few challenges, but it has emerged out to be a suitable solution to maintain secured data transactions in huge IoT networks. The valid transactions prevent theft and spoofing.

5.4 Blockchain in Business

a. Banking industry is relying on the Blockchain model because of its potential of maintaining secured transactions. It provides increased transparency and reduces the costs per transaction by eliminating the presence of any intermediate third party and due to its nature of the distributed and shared database. The ledger being distributed in nature records transactions with enhanced trust levels among communicating parties [4]. Since Blockchain verifies each and every transaction flowing through the IoT network, it does not allow the communicating members to manipulate the transactions, and thus improves security.

Blockchain designates the task of approving transactions to all the members of the network and restricts a single entity to hold the power of controlling the entire system. ICICI bank in India adopted this security model for ensuring secured bank transactions.

b. IBM used a concept termed "ADEPT," a system developed in collaboration with Samsung uses Blockchain technology. IBM "ADEPT" uses the element of Bitcoin underlying design to build a distributed network of devices – a decentralized IoT. This concept of ADEPT (Autonomous Decentralized Peer-to-Peer Telemetry) taps Blockchains to provide the backbone of the system, utilizing a mix of proof-of-work and proof-of-stake to secure transactions. Using the ADEPT system, Samsung Washing machine W9000 uses smart contracts to order detergent supplies by automatic payment and keeps track of its shipment and deliveries. All this information is also shared by the system to the owner.

c. **CISCO JASPER**: Jasper, now part of Cisco, offers a cloud-based software platform for the Internet of Things (IoT) and, more explicitly, to enable product businesses to become IoT service businesses. It integrates the Blockchain technology in its business processes.

CISCO JASPER helps enterprises to enhance their impact on supply chains by using an IoT framework to launch, manage, and monetize the enterprise services. It offers an effective increase in efficiency by a predictive maintenance for performance analytics, and real-time maintenance through the Control Center, thus resulting in better revenues and better customer experiences.

5.5 Benefits of Blockchain

The Blockchain can bridge the missing link to address the issues of privacy, security, and reliability of Internet of Things industry. Blockchain could be used to track billions of network devices and enable the processing of transactions and coordination between devices, thus leading to significant savings to the IoT-based manufacturing industry. The major advantages of Blockchain can be summarized as:

  a. **Public in nature:** Every participating node can view the blocks and also the transactions stored in them. The actual content of transactions is encrypted.
  b. **Secure:** Blockchain uses a ledger which cannot be tampered and cannot be modified by the intruders since the blocks do not exist at the central location. There is no single thread of communication that could be deciphered, hence making the man-in the middle attack difficult to occur [2]. The algorithms that have been implemented prevent cryptographic attacks and accelerate the growth of business by building trust between communicating parties.
  c. **Decentralized:** The right to approve transactions is not held by a single authority. And thus the servers have autonomous capabilities. The decentralized and autonomous capabilities of the Blockchain make it an ideal component to become a foundational element of IoT solutions. There is a single public ledger, and hence the costs and storage demand reduce compared to multiple ledgers.
  d. **Transparency and immutability:** Although any changes in the distributed ledger can be viewed by all the nodes present in the IoT network (transparency), none of them can be deleted by any of the nodes (immutability).
  e. **High-quality data and longevity:** The data output by the Blockchain is accurate, consistent, and available to all participating nodes. This strengthens the data against failure and malicious attacks.

f. Faster transactions: The superior feature of Blockchain is its computing time. Blockchain models the P2P communication by public key cryptography. Centralized databases create more computing overhead.

g. Lower transaction costs: The costs of transactions reduce significantly since there is no intermediate third party.

5.6 Challenges of Blockchain in IoT

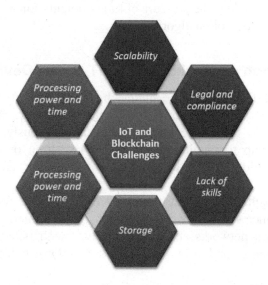

a. Scalability: The ever-growing IoT network might lead to the issue of centralization in the near future, and with the growing pace of time, it might require the management of records which would be a dark cover on the future of this promising technology.

b. Issues of legal and compliance contracts: The current Blockchain models in the IoT environment are not bounded by compliance regulations, but the service providers and IoT manufacturers might face some serious problems since IoT is entirely a new ecosystem. Certain business threats cannot be escaped by the use of Blockchain technology.

c. Lack of skills: The concept of Blockchain is an emerging trend and not much experts are available for its implementation domain. Industries require hiring of teams and proficient engineers for deploying and maintaining the Blockchain model.

d. **Storage:** The Blockchain model drives out the centralized server for storing device ID's and transactions, but the block of valid records or the ledger must be stored at devices only. The subsequent storage of ledgers elevates the size of storage required. Most of the smart devices such as sensors do not have enough storage potential and thus concern for sufficient storage space arises. In the end of 2016, the Blockchain size reached more than 90 GB. Thus, storage is an important disadvantage.

e. **Processing time and power:** The IoT systems being integrated with Blockchain lead to the overhead of processing of algorithms for encryption and thus induce the pressure of time computation at a desired speed for different encryption algorithms.

5.7 Application of Blockchain in IoT Smart Devices

With the aid of Blockchain, the unique history of smart devices can be tracked. The distributed ledger records data exchanges, and Blockchain empowers them to work independently and autonomously. The record of history would help to monitor each and every action of devices. These smart devices could be a laundry machine, a dishwasher, and robo vacuum cleaners.

The smart vehicles can keep a maintenance record for itself. It can have a maintenance schedule, diagnosis record, and payment details incurred on it. Blockchain networks abstract themselves with potential to become independent agents, and refer themselves as "Distributed Autonomous Corporations."

Some of the Blockchains in IoT applications are as follows:

a. **Online shopping**
   Blockchain improves workflow and status of shipments in real time. Online shopping manages and handles many elements on warehouse end as well as on customers end. Contents in container, bills, invoices, carriers, etc., need to be monitored. Blockchain helps buyer and seller parties to establish a transparency. Errors and frauds are reduced. Time spent by the products in shipping process and transit is reduced and management of inventory is improved, thus ultimately reducing cost.

b. **Blockchain in a supply-chain ecosystem**
   The immutable framework of Blockchain is a security-rich and highly transparent network, and provides the participants with an end-to-end visibility. The progress of goods through supply chains can be viewed

by each participant. Bills and invoices can be seen at any time. Modification and deletion of any record are not possible without consensus of involved parties in the network. Thus, Blockchain stimulates a sustainable transport of goods by aggregating shipping processes, and thus offers a trusted access to merchandize.

c. **Blockchain in diamond business**

Blockchain will revolutionize the future of digital assets. Blockchain has the power to run business transactions around the globe by reducing friction and enhancing trust. Diamond mining industry is an example of it.

Diamonds being extremely precious and highly rare face the threats of smuggling, fraud, counterfeit diamonds, and unethical mining of stones. Blockchain in future would enable a more transparent system. Blockchain can be configured for the following processes:

- A photo with high resolution of each diamond can be kept in records.
- Real-time transactions can be maintained for every payment transaction.
- Authentication certificate can be designed and hold by parties involved in transactions.
- The serial number of diamond pieces, their weight, clarity, and carat can be maintained.

5.8 Summary

The IoT ecosystem needs to coordinate and collaborate with network connectivity by keeping the concerns of time, cost, and security. In this chapter, we have discussed Blockchain as a model to provide the secure data exchanges among connected IoT devices. All the devices that have been integrated must comply with the processing of the Blockchain model along with the network infrastructure and IoT framework. The Blockchain is a good mechanism to preserve the data integrity in an IoT network. With the enhancing domain of IoT technologies, Blockchain ensures that smart devices get autonomous functioning. The Blockchain model in the IoT framework enhances the security aspects for device authentication and data verification, and encrypts and verifies valid transactions at all levels. This way, the IoT framework can be secured and safeguarded with the Blockchain if a few shortcomings of it are overcome.

References

[1] Di Francesco Maesa, D., Marino, A., & Ricci, L., "Uncovering the Bitcoin Blockchain: an analysis of the full users graph," *2016 IEEE International Conference on Data Science and Advanced Analytics*, Montreal, QC.

[2] Available at: http://www.cnbc.com/2016/10/22/ddos-attack-sophisti cated-highly-distributed-involved-millions-of-ip-addresses-dyn.html

[3] Available at: http://www.cio.com/article/3027522/internet-of-things/ beyond-bitcoin-can-the-Blockchain-power-industrial-iot.html

[4] Available at: http://digital-library.theiet.org/content/journals/10.1049/et. 2016.1003

6

Cryptic Mining for AVK-based IoT Cryptosystem and Client-side Encryption Perspective

Shaligram Prajapat

International Institute of Professional Studies, Madhya Pradesh, India
Devi Ahilya University, Indore, Madhya Pradesh, India

Automatic variable key (AVK)-based enciphering schemes have claimed to be energy efficient for communication and information exchange among IoT-based devices. This chapter presents parametric versions of symmetric key-based encryption algorithms. The parametric version of AVK emphasizes on generation and usage of key solely based on the parameter. The common method of key construction relies on numeric keys. Here, the key construction process has been extended for generation of alphanumeric keys and a domain of parameters (selected based on user's personal information). It is very common to use numeric or alphabetic series, recurrence relations, or location information which can jeopardize the safety of a system. The cryptic model gives freedom to the user for the parameter selection and variation in the parameters. This chapter presents the analytics perspective of the parametric model in the light of association rule mining for the AVK-based symmetric cryptosystem. Useful inferences and results from testing of cryptic association rule mining support the view of auditing encryption algorithm and identify the power of using a large number of parameters for secure information exchange among IoT devices.

6.1 Introduction

With the boom of the Internet of things (IoT), cybercrime will not only be limited to gaining information of credit card details and database information associated with it, but also penetration to items of real life and the human body. IoT has enabled computer moving on the wheels (self-driving cars) and flying computing devices (smart aero-gadgets and planes) to scanners (MRI) and insulin pumps. However, it is difficult to estimate and analyze how secure these instruments are compared to stand-alone/connected stationary computers. There are fair chances that cybercriminals and hackers may control the human-less cars and even pacemakers from remote locations. By proactive alertness, such security issues can be improved. Apart from IoT perspectives of information security, the cloud-based security issue is also a growing concern. The rationale for considering cloud perspective is that IoT data are stored and processed in the cloud that provides the benefit of agility, storage capacity, performance, and high availability. The third party providing this service needs to explore secure storage of this massive stored information. The protection of data on to the cloud is not an easy job. Many organizations protect the data that live in traditional on-site data centers, but with cloud-based IoT information repository, there are a different set of challenges due to dependence on the cloud.

The level of responsibility and degree of information to be secured are a big challenge together with the following issues in storing and retrieving from the available alternatives:

1. *Lack of resource utilization:* When organizations move only non-critical data to the cloud, the potential of the cloud is underutilized and it limits the growth of business.
2. *Application of existing solutions to the cloud:* By using intelligent application of the encryption process across the cloud infrastructure, data can be secured. Organizations take their current data center security solution and apply them to the cloud. The data stored on the cloud are shared by multiple organizations and managed by administrators or employees of the organizations. However, the sharing of cloud infrastructure may have several issues like: Data governance, regulatory, and compliance issues. These solutions focus on parameter security and access control. They do not protect the data itself. With multi-tenancy in the cloud, it needs radically different concepts of parameters and access control for protection of data. Cloud data protection is considered to be the biggest roadblocks for organizations planning to the cloud. Using efficient encryption techniques is the only choice for securing your data in

cloud, because your data are kept on a machine that is shared by multiple machines. It significantly reduces the risk associated with many data governance and regulatory compliance issues.

3. *Prevention and risk mitigation*: An organization must follow the necessary measures to prevent the loss of information and must be prepared to fortify its security against malicious attempts to steal organization's confidential information.

4. *Understand what is being shared and who is taking care of* information security at cloud: The organization must have full knowledge and vision about facilities from cloud service providers.

Besides the above four points for securing information, *encryption at end-user-service* is also another alternative. The encryption at the user end is the approach of encoding data at the origin (user end), and then it is sent to the cloud storage provider. The encryption key of the user is not available to the cloud service provider, making it hard or impossible for cloud storage provider to decipher data at the cloud. This schema allows the maintaining zero-knowledge system whose host is unable to access the data, and ensures a high level of privacy and exceptionally robust data security strategy. It eliminates the secret data to be viewed by third entity including service provider, and client-side encryption ensures that data and files that are stored in the cloud can be viewed only on the client side of the exchange. This prevents data loss and the unauthorized disclosure of private or personal files, providing increased peace of mind for both personal and business users. Thus, confidentiality, integrity, and authenticity of information can be achieved to protect IoT data. Recently, many organizations are providing this facility (Confidentiality, Integrity and Authenticity of information). As of the beginning of 2016, none of the big-giants are providing client-side encryption.

The biggest challenge of leaving security at the client side or end-user side is mishandling of keys, poor selection, and choice of keys or passwords. An example of a poor key would be typically having a key like "1234abcd" or date of birth or "qwerty" or a similar random password chosen by users that risks the safety of the system. It basically results in choosing a key from information that is available from public–social domain, or easy to guess.

6.2 The Classical Model of Automatic Variable Key (AVK)

"Perfect Secrecy" is the requirement of a cryptosystem, where after a cryptogram is attacked by an intruder, the posterior probability of this cryptogram is the same as the prior probability of the same messages before the interception [1, 2]. It reveals that perfect secrecy is surely achievable but the condition is that if the number of messages is finite, then the same number of possible keys should be there, or if the message is thought of being constantly generated at a given rate, the key must be generated at the same or at a higher rate [2–5].

Secured transport of information over the network is a pertaining research challenge in today's context. The problem continues to aggregate with an increasing volume of network traffic, which is evident from several research studies [12, 14, 15]. Security basically refers to the protection of the data against intentional modification, loss or damage, and fabrication of data and/or deliberate disclosure of data to unauthorized persons or miscreants. In the classical paper on Information Security, Shannon, the father of Information theory has already established that perfect security can be achieved only when the key is made to vary from session to session and/or data to data [6, 7].

6.3 Mechanism of Variation in Key of Automatic Variable Key (AVK) Framework

Realization of the time variable key is difficult to achieve, as a key must be communicated between the sender and the receiver [1, 3] from time-to-time. The automatic variable key (AVK) for a session between Alice and Bob is to be sent initially as K_0, and they exchange data D_0. The key is now variable, and after every transmission, it changes dynamically such that: with initial key K_0 *initial secret key*, the future keys can be constructed by $K_i = K_{i-1} \oplus D_{i-1}, \forall i > 0$, where D_{i-1} and K_{i-1} *are data and key of* $(i-1)th$ *session.*

The variable key if implemented as suggested will not result in the repetition of patterns, unlike in the normal mode. AVK depends on the data sent previously. There is no guarantee that the previously sent data may not be stolen. This needs further investigations for proper utilization of the time-variant key [2, 4]. In order to solve the problem, there should be some technique such that the previously sent data are protected or there are some storage media where a replica of the data can be stored. The circuit will behave as a multiplexer where the control signal will govern whether there is a need to swap-in of data from the data-store or not [6–9].

Designing a good cryptosystem not only requires encryption algorithms and key management protocols but also requires necessary testing to test the design under cryptic pattern discovery and mining.

The parametric Fibo-Q model uses parameter n as input for computation of key. Sparse approach uses location (i, j) as two parameters for computation of keys. The keys used in these approaches were numeric keys. To extend the idea over alphanumeric keys, the parameters like

p_1 = part of vehicle number, p_2 = nickname, and p_3 = date of birth of spouse, can be used. The possible key samples may be *mp09t!nku020284*, *mp09b0b020284, mp09t!nku020286,* etc. The keys generated in this fashion are $\{P_1P_2P_3, P_1P'_2P_3, P_1P_2P'_3, P'_1P_2P'_{3...}\}$. Hence, by changing parameter values, variable keys can be generated. The alphanumeric mixed key construction using this approach is to be analyzed from the hacker's perspective. The parameters from the possible parameter set used in different sessions can be found through association rules. The parameters may be associated with the device location of IoT and used to generate its secret key for secure communication. The keys may also be used as OTP for authentication. To simplify the situation, we have taken real-world persons as a device and associated information of the person will act as a parameter for the generation of a secret key. The public information of element of the IoT network may be captured by peeping tom and used to predict the parameters and also the final key. The next section deals with basic association rules and examples and presents mining and frequent patterns discovery for the proposed parametric AVK model analysis.

6.3.1 Apriori Approach for Parameter Prediction

Conventionally, the association rule $X \rightarrow Y$ indicates that if the key (antecedent) appears, then the parameter set (consequent) $p_i, p_j, \ldots, p_k$ also tends (with highly probable) to appear, where X and Y may be single parameters or set of parameters (in which the same parameter does not appear in both sets) [10]. In other words, X and Y would be found together frequently in the given training set and they do not show a causal relationship [11].

For example, the number of parameters in session Table 6.1 is 16. The key of a particular session is constituted from a variable number of parameter terms $\{p_i, p_j, \ldots, p_k\}$ and it is denoted by $f(p_i, p_j, \ldots, p_k)$, where $p_i, p_j, \ldots, p_k$ are variables specific to a particular session. Further, assume that information of n-sessions is available (Table 6.1). Each session of Table 6.1 is denoted by $S = \{S_i, S_j, \ldots, S_k\}$ with a unique session-Id and function f(.) with a set of parameters (possibly a small subset) constituting each session key. Each session key of m parameters is with key $f(p_i, p_j, \ldots, p_k)$. Typically, the session key is varied due to differences in

Table 6.1   Session-wise parameters of a key

S_i	Session Key	Parameters Used
S_1	$f(p_1, p_2, p_4, p_6, p_{16})$	$p_1, p_2, p_4, p_6, p_{16}$
S_2	$f(p_1, p_3, p_4, p_6)$	p_1, p_3, p_4, p_6
S_3	$f(p_4, p_5, p_7, p_9, p_{10})$	$p_4, p_5, p_7, p_9, p_{10}$
S_4	$f(p_2, p_4, p_6, p_3, p_9)$	p_2, p_4, p_6, p_3, p_9
S_5	$f(p_2, p_3, p_5, p_7, p_9)$	p_2, p_3, p_5, p_7, p_9
S_6	$f(p_{10}, p_{15})$	p_{10}, p_{15}
S_7	$f(p_1, p_2, p_4, p_6, p_{10})$	$p_1, p_2, p_4, p_6, p_{10}$
S_8	$f(p_8, p_{10}, p_{15})$	p_8, p_{10}, p_{15}
S_9	$f(p_2, p_3, p_4, p_5, p_6)$	p_2, p_3, p_4, p_5, p_6
S_{10}	$f(p_2, p_3, p_5, p_7, p_9)$	p_2, p_3, p_5, p_7, p_9
S_{11}	$f(p_2, p_4, p_9)$	p_2, p_4, p_9
S_{12}	$f(p_2, p_4, p_6, p_7, p_9)$	p_2, p_4, p_6, p_7, p_9
S_{13}	$f(p_1, p_2, p_3)$	p_1, p_2, p_3
S_{14}	$f(p_3, p_4, p_5, p_7, p_9)$	p_3, p_4, p_5, p_7, p_9
S_{15}	$f(p_5, p_6)$	p_5, p_6
S_{16}	$f(p_7)$	p_7
S_{17}	$f(p_7, p_8, p_9)$	p_7, p_8, p_9
S_{18}	$f(p_1, p_2, p_4 \, p_6)$	$p_1, p_2, p_4 \, p_6$
S_{19}	$f(p_2, p_3, p_5, p_7, p_9)$	p_2, p_3, p_5, p_7, p_9
S_{20}	$f(p_4, p_5, p_7, p_9)$	p_4, p_5, p_7, p_9
S_{21}	$f(p_{10}, p_{15}, p_{16})$	p_{10}, p_{15}, p_{16}
S_{22}	$f(p_2, p_3, p_4, p_6)$	p_2, p_3, p_4, p_6
S_{23}	$f(p_5, p_7, p_9, p_{10}, p_{11})$	$p_5, p_7, p_9, p_{10}, p_{11}$
S_{24}	$f(p_{11}, p_{12}, p_{13})$	p_{11}, p_{12}, p_{13}
S_{25}	$f(p_{13}, p_{14}, p_{15})$	p_{13}, p_{14}, p_{15}

the number of parameters. A cryptanalyst has a record of what parameters are used for each session to generate its session key. The goal of cryptanalyst here is to find association relationships from a given large number of session keys, to identify the parameters that tend to occur together. In Table 6.1, each row shows the set of parameters that are used in each session.

The cryptanalyst analyzes session keys shown in Table 6.1 to identify which parameter sets are used frequently in a session. Let p_6 and p_9 are the two parameters that are used together frequently, then the hacker may start predicting by having one parameter information, in the hope that the second parameter information can be found by the obtained association rule. Given a large set of transactions, a procedure is needed to discover all association rules, such that all rules satisfying the desired constraints are found in an efficient manner. Out of these rules, only practical or actionable rules are important.

6.3.2 Phase-1: Computation of Frequent Set

Let S be a transaction log with information about the 25 sessions, i.e., with session keys using parameters from parameter space of 16 possibilities, i.e., $P = \{p_1, p_2, \ldots, p_{16}\}$, where each session key is generated by a random selection of some parameters and secret key generation algorithms. In the AVK environment, it is assumed that a cryptanalyst or hacker somehow recorded traces of parameters used in a few sessions, say 25, without the information of function, and a cryptanalyst may be interested in finding out the frequent set of parameters or in guessing future parameters based on the association rules that can be used to predict the future session key.

The frequency of each parameter in the session logs is given in the following set, where each set element = {parameter, frequency of parameter} is listed below:

$\{\{p_1{:}4\}, \{p_2{:}13\}, \{p_3{:}10\}, \{p_4{:}11\}, \{p_5{:}9\}, \{p_6{:}9\}, \{p_7{:}10\}, \{p_8{:}2\},$
$\{p_9{:}11\}, \{p_{10}{:}6\}, \{p_{11}{:}2\}, \{p_{12}{:}1\}, \{p_{13}{:}2\}, \{p_{14}{:}1\}, \{p_{15}{:}4\}, \{p_{16}{:}2\}\}.$

Assume support of parameters (25% support in 25 sessions) to occur in at least seven sessions for computing the first frequent parameter set L_1 in Table 6.2.

Computation of C_2: There are 21 candidates for the two-parameter set of C_2 $\{(p_2, p_3), (p_2, p_4), (p_2, p_5), (p_2, p_6), (p_2, p_7), (p_2, p_9), (p_3, p_4), (p_3, p_5),$
$(p_3, p_6), (p_3, p_7), (p_3, p_9), (p_4, p_5), (p_4, p_6), (p_4, p_7), (p_4, p_9), (p_5, p_6), (p_5,$
$p_7), (p_5, p_9), (p_6, p_7), (p_6, p_9), (p_7, p_9)\}$. Their frequency of items is reported in Table 6.3.

The two-frequent set and the candidate set of three parameters with the respective frequency are given in Tables 6.4 and 6.5, respectively. The three-frequent parameters are shown in Table 6.6.

6.3.3 Phase-2: Computation of Association Rule1

The three-frequent-parameter set is computed from L_2. Taking one parameter in antecedence from $\{p_2, p_4, p_6\}$ will result in

$$\{p_2 \rightarrow p_4, p_6; p_4 \rightarrow p_2, p_6; p_6 \rightarrow p_2, p_4\}$$

Table 6.2 L_1: First frequent parameter set

Parameter	p_2	p_3	p_4	p_5	p_6	p_7	p_9
Frequency	13	10	11	9	9	10	11

Table 6.3   Two-frequent-candidate set C_2

Parameter Sets	Frequency
(p_2, p_3)	9
(p_2, p_4)	8
(p_2, p_5)	4
(p_2, p_6)	8
(p_2, p_7)	4
(p_2, p_9)	6
(p_3, p_4)	5
(p_3, p_5)	4
(p_3, p_6)	5
(p_3, p_7)	4
(p_3, p_9)	6
(p_4, p_5)	4
(p_4, p_6)	9
(p_4, p_7)	3
(p_4, p_9)	4
(p_5, p_6)	1
(p_5, p_7)	7
(p_5, p_9)	7
(p_6, p_7)	1
(p_6, p_9)	2
(p_7, p_9)	9

Table 6.4 L_2: The two-frequent parameter set

(p_2, p_3)	9
(p_2, p_4)	8
(p_2, p_6)	8
(p_4, p_6)	9
(p_5, p_7)	7
(p_5, p_9)	7
(p_7, p_9)	9

Table 6.5 C_3 – Candidate sets of three-parameter set and frequency

The Candidate Set – Three-Parameter Set	Frequency
p_2, p_3, p_4	4
p_2, p_3, p_6	4
p_2, p_4, p_6	8
p_5, p_7, p_9	7

Table 6.6 L_3 – Three-frequent-parameter set

Three-Frequent-Parameter Set	Frequency
p_2, p_4, p_6	8
p_5, p_7, p_9	7

Rules with the two-parameter set in antecedence position are

$$\{p_4, p_6 \to p_2; p_2, p_6 \to p_4; p_2; p_4 \to p_6\}$$

Taking support of 8 as computation of confidence, association rules for parameters p_2, p_4, and p_6 are given in Table 6.7.

With support of 7 as computation of confidence for p_5, p_7, and p_9, association rules for parameters are shown in Table 6.8.

With confidence $= 70\%$, a cryptanalyst or hacker may infer all seven rules (except rule number 3). The generated rules are: p4→p2, p4→p6, p6→ p2, p6→p4, p4, p6→p2, p2, p6→p4, p2, p4→p6, p5→p7, p5→ p9, p7→p5, p7→p9, p7, p9→p5, p5, p9→p7, p5, p7→p9, p2→p3, and p3→p2 (Note that the rules have been decomposed like p4→p2, p6 by two rules p4→p2 and p4→p6).

Table 6.7 Association rules for p2, p4, and p6

Rule	Support of (p_2, p_4, p_6)	Frequency of Antecedence	Confidence (%)
$p_2 \to p_4, p_6$	8	13	0.61%
$p_4 \to p_2, p_6$	8	11	0.72%
$p_6 \to p_2, p_4$	8	9	0.89%
$p_4, p_6 \to p_2$	8	9	0.89%
$p_2, p_6, \to p_4$	8	8	1%
$p_2, p_4, \to p_6$	8	8	1%

Table 6.8 Association rules for p5, p7, p9

Rule	Support	Frequency of Antecedent	Confidence (%)
$p_5 \to p_7, p_9$	7.0	9	0.78%
$p_7 \to p_5, p_9$	7.0	10	0.7%
$p_9 \to p_5, p_7$	7.0	11	0.64%
$p_7, p_9 \to p_5$	7.0	9	0.78%
$p_5, p_9 \to p_7$	7.0	7	1%
$p_5, p_7 \to p_9$	7.0	7	1%

6.4 Experimental Analysis of Parametric Cryptosystem

For the extended parametric AVK model (with key variation by changing parameters), the objective of the cryptosystem designer is to make it as difficult as possible for a hacker to make informed guesses about the chosen parameters or keys. Thus, there is no alternative but a brute-force search, trying every possible combination of letters, numbers, and punctuation. A search of this sort, even conducted on a machine that could try one million keys per second (most machines can try less than 100 per second), requires, on an average, over 100 years to complete. The ease of method is also essential that is possible only when the parameters are taken from the neighborhood of the user. With this as a goal, and using the information in the preceding text, a survey of patterns used for key formation has been discussed in subsequent sections. It is better to reconsider the features of AVK model before going to the next section.

6.4.1 Features of Automatic Variable Key

1. Fix the key with an optimum size and change the key from session to session. In the parametric AVK version, it can be achieved by varying the parameters.
2. Additional level of security can be implemented by sharing only parameters on a public network, instead of exchanging keys.
3. By changing parameters, the session keys (AVK) are generated. A new key is considered by changing some parameters in the previous key.

$K_0 \leftarrow Initial\ Key$

$K_{i+1} \leftarrow K_i \odot \{p_i, p_j, p_k \ldots\}$

Where $\odot$ represents concatenation or manipulation in some parameters

$p_i, p_j, p_k \ldots$ *of key K_i to construct K_{i+1}*

In Fibo-Q approach, a key is varied with parameter $= n$ along with computation of $f(n)$, $f(n-1)$ and $f(n+1)$. It uses any number of parameters between 1 to n for the next key if $n < 35$. Otherwise choose random n with a mutual agreement by the sender and the receiver. In Sparse approach, a key is varied with location parameters $p_{ij} =$ location (i, j) (This approach is applicable for moving devices, where the next location coordinates will produce the next key).

In general, for a cryptosystem S, if parameters are from a domain set $p = \{p_1, p_2, \ldots, p_n\}$ and by choosing the number of parameters to a minimum 3 to maximum 7 keys can be constructed and shared among communicating parties. Since parameters are exchanged over a public network and may be found or recorded on the log, the security of the parameterized model can be tested with mining techniques. For testing of the parametric model using mining techniques, the following information has been collected from a public domain (e.g., a department of university). The broad view of responses has been highlighted as suggestions to the cryptic system designer for the parametric model.

1. The system is secure against association rule mining, even though the parameters are public.
2. The model generates huge frequent sets that are hard to determine the future parameters for sessions.
3. The key construction policy may help to suggest future keys/parameters for AVK using a convenient approach.

6.4.2 Experimental Setup for Association Rule Analysis

In the data collection step, the online survey questions are designed to test the parameter usage and analyze the behavior of the changing parameter. The following list of 23 questions is asked to know about potential parameters which are used commonly by users (* indicates mandatory response):

1. Do you use your first or last name in a form (nickname)?*
2. Do you use your DOB, anniversary date, etc.? *
3. Do you use your login name in any form (as-is, reversed, capitalized, doubled, etc.)?*
4. Do you use your spouse's or child's name?*
5. Do you use other information which can be easily obtained about you, such as license plate numbers and house numbers?*
6. Do you use PAN number, Aadhar number, Social Security Number, Passport number, etc.?*
7. Do you use telephone numbers?*
8. Do you use brand of your automobile and vehicle number?*
9. Do you use a password that consists of all digits, or all alphabets?*
10. Do you use a word from language dictionaries, spelling lists, or other lists of words?*
11. Do you use a password shorter than six characters?*
12. Do you use a password with mixed-case alphabets?*
13. Do you use a password with no alphabetic characters, e.g., digits or punctuation?*

14. Do you use a password that is easy to remember, so you don't have to write it down?*
15. Do you use a password that you can type quickly, without having to look at the keyboard? This makes it harder for someone to steal your password by watching over your shoulder?*
16. Do you use DOB of your girlfriend/boyfriend/spouse?*
17. Do you use name of your favorite leader/actor/movie/game in password?*
18. Do you use station code/country code/area code?*
19. Do you use your institute name/office name (in some form)?*
20. Do you use the first letter of each word from a line of a song/book or poem?*
21. Do you use your residential address, city name, state name, country name, etc.?*
22. Do you use your ID numbers provided by different institutions such as roll number, subscription number, or exam id?*
23. Do you use any other type of personal information for constructing a field/parameter of password/key?

The responses of these questions may be taken as a log of information for prediction of parameters. In real-world parameters, we have received responses as a probable parameter for key generation on a five-point Likert scale. For sample size $N = 100$, the responses are considered in the format: (*I never use it*)1 2 3 4 5 (*I always use it*)

The parameter set is denoted in subscripted notations by $P = \{p_1, p_2, \ldots, p_{22}\}$

The parameters, p_1–p_{22}, used by common users for securing parametric communication over public network for key constructions include the following:

p_1	$=$	First name/last name/nickname/alias
p_2	$=$	DOB/Anniversary date
p_3	$=$	Public name in any form (as-is, reversed, capitalized, doubled, etc.)
p_4	$=$	Do you use your spouse's or child's name?
p_5	$=$	Information easily obtained about you (license plate numbers, house numbers)
p_6	$=$	PAN number/Aadhar number/Social Security Number/Passport number

(Continued)

p_7	=	Telephone numbers/mobile numbers
p_8	=	Automobile and vehicle numbers
p_9	=	Password of all digits/all the same letters
p_{10}	=	Word contained in English or foreign language dictionaries/spelling lists/other lists of words
p_{11}	=	A password shorter than six characters
p_{12}	=	Passwords with mixed-case alphabetic
p_{13}	=	No alphabetic characters, e.g., digits or punctuation
p_{14}	=	Key that is easy to remember, so you don't have to write it down
p_{15}	=	A key that you can type quickly, without having to look at the keyboard (making harder for someone to steal your password by watching over your shoulder)
p_{16}	=	DOB of your girlfriend/boyfriend/spouse
p_{17}	=	Station code/country code/area code
p_{18}	=	Current institute name/office name (in some form)
p_{19}	=	Use of the first letter of each word from a line of a song/book or poem
p_{20}	=	Part of residential address/city name/state name/country name
p_{21}	=	ID numbers provided by different institutions/organizations such as roll numbers, subscription numbers, and exam id
p_{22}	=	Non-personal information for constructing a field/parameter of a password/key

The responses received are collected through Google form, which was shared over professional network (LinkedIn), social network (Facebook and Google plus), and Google groups of aluminous of DAVV to get real trends. The baseline from the hacker's perspective on the parametric AVK model may be used to find out which parameters are favorable for key constructions. Which among those are the most prominent or frequent? Is there any association among these parameters? But with a large number of pairs, determination of correlation will be costly. Finding correlation will be easy in case of less number of parameters. We used the SPSS tool to analyze the survey result, because it provides grouping of parameter sets into factors, and allows finding out the favorable parameters for key construction. The result shown in Table 6.9 presents the probability of making choices of parameter sets. The tool also provides correlation among these parameters,

Table 6.9 Group of frequent set of related parameters with probability

Frequent Parameters Used	% of Time Choice has been Made	Choice/Preference
p_6, p_7, p_8	19.76%	First
p_{19}, p_{20}	13.90%	Second
p_4, p_{15}, p_{18}	10.68%	Third
p_{10}, p_{11}, p_{13}	09.21%	Fourth
p_9, p_{17}	07.87%	Fifth

but in case of more parameters, association rule generation requires less efforts for inference. Thus, in case of large number of parameter sets, mining algorithms are suitable. Mining over a large number of responses out of these 22 parameters, a group of favorable parameters for key construction with respective probability of the frequent parameters are identified as mentioned in Table 6.9.

6.4.3 Frequent Patterns Generated for Parametric AVK Model

Traditional Apriori algorithm is applied on response data with a variable number of parameters. To analyze the responses, Apriori algorithm is applied to response data with tab delimited text file format used as the input file. The same file also stores the end results. The algorithm can be used in two ways: first, find the frequent parameter set by pruning the candidate parameter set with a user-defined support threshold value. Later for rule generation, the minimum support and confidence are varied over range (1–100%). With respect to variable support, variation in the number of frequent parameters, the time consumed in seconds for generation of association rules, and corresponding file size has been captured as shown in Table 6.10.

6.5 Analysis of Cryptic Mining Results of Parametric Model

Analysis of observation (Table 6.9) inferences can be drawn as: the cardinality of 22 parameters and 100 survey responses of the frequent patterns are too large in number. For the lower support less than 5%, frequent patterns range from 70,000 to 63,00,000 which indicates that the frequent combinations of parameter choice by few people are too big.

1. In reality, it is true that the possibility of choosing the parameters in the same way by different people is very low.
2. The result also shows that as we increase the support percentage, the frequent patterns are decreasing dramatically.

Table 6.10   Frequent set generated with various support values

S. No.	Support	No. of Frequent Items	Time Taken	File Size
1	1%	6,316,071	3.13 s	303 MB
2	2%	2,725,378	1.26 s	123 MB
3	3%	675,419	0.35 s	28 MB
4	4%	70,988	0.04 s	2.31 MB
5	5%	18,558	0.01 s	555 KB
6	6%	7080	<0.01 s	192 KB
7	7%	3405	<0.01 s	45 KB
8	8%	1891	<0.01 s	27 KB
9	9%	1178	<0.01 s	18 KB
10	10%	793	<0.01 s	12 KB
11	11%	530	<0.01 s	9 KB
12	12%	405	<0.01 s	7 KB
13	13%	323	<0.01 s	6 KB
14	14%	263	<0.01 s	5 KB
15	15%	227	<0.01 s	2 KB
16	20%	98	<0.01 s	649 Bytes
17	30%	35	<0.01 s	337 Bytes
18	40%	18	<0.01 s	241 Bytes
19	50%	13	<0.01 s	116 Bytes
20	60%	07	<0.01 s	64 Bytes
21	70%	04	<0.01 s	15 Bytes
22	80%	01	<0.01 s	15 Bytes
23	90%	01	<0.01 s	15 Bytes
24	100%	No frequent item	<0.01 s	0 Bytes

3. Frequent patterns obtained from higher support are very few in numbers and may be analyzed easily, but at the same time, it is observed that frequent patterns are obvious in nature and their size (item set size) is also small. So, the generic parameters do not require analysis. For example, in our case, parameter 12 (selection of mixed case for key) is very natural.

4. Another conclusion can be reached based on the result obtained (see frequent patterns in Table 6.10) that the frequent patterns with higher supports are of no use because there are rare chances that many people think in the same way for parameter selection of keys. But a few people may think in the same way, and for that, we need to decrease the support percentage and then patterns increase exponentially in varying size, which makes it impossible to analyze it manually and very hard for a system to breach it.

Association rules from the above frequent parameters are generated, and Table 6.11 shows some of the discovered rules after analysis.

Table 6.11 Association rule with different support and confidence

S. No.	Support (%)	Confidence (%)	Time (s)	File Size	No. of Rules
1	1	80	108.27	3.5 GB	62,151,312
2	1	90	70.94	3.5 GB	62,050,814
3	1	100	71.91	3.5 GB	62,050,212
4	2	80	27.21	1.37 GB	25,389,899
5	2	90	29.34	1.36 GB	25,289,401
6	2	100	28.41	1.36 GB	25,288,799
7	3	80	5.67	288 MB	5,619,616
8	3	90	5.63	284 MB	5,519,118
9	3	100	5.62	284 MB	5,518,516
10	4	80	0.46	170.4 MB	407,512
11	4	90	0.30	13.3 MB	307,014
12	4	100	0.30	13.3 MB	306,412
13	5	80	0.15	5.92 MB	145,504
14	5	90	0.09	1.82 MB	45,006
15	5	100	0.06	1.80 MB	44,404
16	10	80	0.02	67.4 KB	1995
17	10	90	0.02	24.5 KB	734
18	10	100	<0.01 s	4.24 KB	132
19	20	80	<0.01 s	3.79 KB	118
20	20	90	<0.01 s	1.10 KB	35
21	20	100	<0.01 s	0 KB	NIL
22	30	80	<0.01 s	1.24 KB	39
23	30	90	<0.01 s	445 Bytes	14
24	30	100	<0.01 s	NIL	NIL
25	40	80	<0.01 s	477 Bytes	15
26	40	90	<0.01 s	285 Bytes	9
27	50	80	<0.01 s	346 Bytes	11
28	50	90	<0.01 s	216 Bytes	07
29	60	80	<0.01 s	208 Bytes	07
30	60	90	<0.01 s	112 Bytes	04
31	70	80	<0.01 s	81 Bytes	03
32	70	90	<0.01 s	81 Bytes	03
33	80	80	<0.01 s	24 Bytes	01
34	80	90	<0.01 s	24 Bytes	01
35	90	80	<0.01 s	24 Bytes	01
36	90	90	<0.01 s	24 Bytes	01

Experimental results discussed in category-A will analyze the effect on changing file size of five parameters with fixing support at 80% (later on to 10, 20, 15, 20, and 22 parameters), later on support is changed to 90% and 100%. The detailed observations are shown in Table 6.12 and Figure 6.1.

Table 6.12 Effect of variation in parameter size (from 5, 10, 15, 20, and 22) with variable support (at confidence – 80%, 90%, and 100%) and corresponding ARM file size [Category-A]

Support (%)	1	2	3	4	5	10	20	30	40	50	60	70	80	90	100
File size (5/80%)	440	441	442	443	444	445	446	447	448	449	450	451	452	453	454
File size (10/80%)	100,966.4	100,966.4	75,264	41,369.6	28,876.8	2385.92	329	120	44	19	19	19	19	19	19
File size (15/80%)	8,514,437	8,514,437	6,375,342	2,904,556	1,226,834	52,633.6	6410.24	1832.96	577	460	326	297	95	95	69
File size (20/80%)	5.04E+08	5.04E+08	2.79E+08	44,774,195	10,695,475	158,875	11,366.4	3225.6	1392.64	1126.4	648	297	95	95	69
File size (22/80%)	3.76E+09	1.47E+09	3.02E+08	1.79E+08	6,207,570	69,017.6	3880.96	1269.76	477	346	208	81	24	24	0
File size (5/90%)	347	407	407	407	347	257	173	69	44	19	19	19	19	19	19
File size (10/90%)	83,968	83,968	58,368	24,473.6	11,878.4	1679.36	329	120	44	19	19	19	19	19	19
File size (15/90%)	7,790,920	7,790,920	5,651,825	2,181,038	512,000	28,160	3543.04	1054.72	577	460	326	297	95	95	69
File size (20/90%)	4.97E+08	4.97E+08	2.72E+08	37,748,736	3,649,044	78,131.2	5744.64	1832.96	1116.16	855	447	297	95	95	69
File size (22/90%)	3.76E+09	1.46E+09	2.98E+08	13,946,061	1,908,408	25,088	1126.4	445	285	216	112	81	24	24	0
File size (5/100%)	440	440	407	407	347	257	173	69	44	19	19	19	19	19	19
File size (10/100%)	83,660.8	93,900.8	58,060.8	24,166.4	11,571.2	1382.4	329	69	44	19	19	19	19	19	19
File size (15/100%)	7,769,948	7,769,948	8,787,067	2,170,552	498,688	14,540.8	1914.88	616	272	219	156	127	45	45	19
File size (20/100%)	4.97E+08	4.97E+08	2.72E+08	37,748,736	3,607,101	35,737.6	2949.12	965	528	403	212	127	95	95	19
File size (22/100%)	3.76E+09	1.46E+09	2.98E+08	13,946,061	1,887,437	4341.76	0	0	0	0	0	0	0	0	0

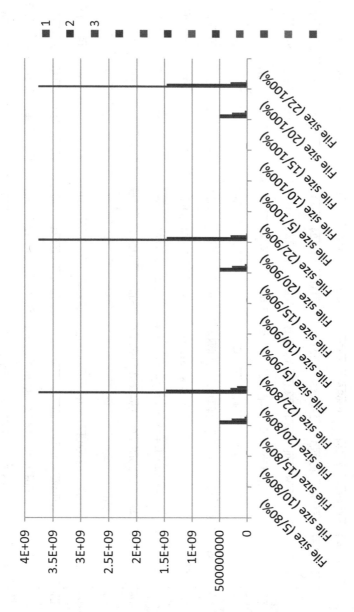

Figure 6.1 Comparative graph of variation in parameter (from 5, 10, 15, 20, and 22) with variable support (at confidence – 80%, 90%, and 100%) and corresponding ARM file size.

Experimental results discussed in category-B will be analyzed over the variable support values from the effect on changing Association Rule Mined (ARM) File Size of five parameters with fixing support at 80% (later on to 10, 20, 15, 20, and 22 parameters); later on, support is changed to 90% and 100%. The detailed observations are shown in Table 6.13 and Figure 6.2.

Experimental results discussed in category-C will be analyzed over the time required variable support values from the effect on changing time for Association Rule Mined (ARM) file size of five parameters with fixing support at 80% (later on to 10, 20, 15, 20, and 22 parameters); later on, support is changed to 90% and 100%. The detailed observations are shown in Table 6.14 and Figure 6.3.

The association rules based on various parameters with support (10%, 20%, 30%...100%) and confidence (80%, 90%, and 100%), and some are discussed in detail.

1. Support 90%, confidence 90% number of rules = 1
   (only parameter P_{12})
2. Support 90%, confidence 80% number of rules = 1
   (Only parameter P_{12})
3. Support 80%, confidence 90% number of rules = 1
   (Only parameter P_{12})
4. Support 80%, confidence 80% number of rules = 1
   (Only parameter P_{12})
5. Support 70%, confidence 90% number of rules = 3
   (Rule-1 parameter P_{12}, rule-2 P_{12} associated with P_{14} and rule-3 parameter P_{12} associated with P_{13})
6. Support 70%, confidence 80% number of rules = 3
   (Rule-1 parameter P_{12}, rule-2 P_{12} associated with P_{14} and rule-3 parameter P_{12} associated with p13)
7. Support 60%, confidence 90% number of rules = 4
   (P_{12}, $P_{12} \leftarrow P_{15}$, $P_{12} \leftarrow P_{14}$, $P_{12} \leftarrow P_{13}$)
8. Support 60%, confidence 80% number of rules = 7
   (P_{12}, $P_{14} \leftarrow P_{15}$, $P_{13} \leftarrow P_{15}$, $P_{12} \leftarrow P_{15}$, $P_{12} \leftarrow P_{14}$, $P_{12} \leftarrow P_{13}$, $P_{14} \leftarrow P_{15} P_{12}$)
9. Support 50%, confidence 90% number of rules = 7
   (P_{12}, $P_{12} \leftarrow P_{15}$, $P_{12} \leftarrow P_{14}$, $P_{12} \leftarrow P_{13}$, $P_{12} \leftarrow P_{15} P_{14}$, $P_{12} \leftarrow P_{15} P_{13}$, $P_{12} \leftarrow P_{14} P_{13}$)
10. Support 50%, confidence 80% number of rules = 7
    (P_{12}, $P_{12} \leftarrow P_{15}$, $P_{12} \leftarrow P_{14}$, $P_{12} \leftarrow P_{13}$, $P_{12} \leftarrow P_{15} P_{14}$, $P_{12} \leftarrow P_{15} P_{13}$, $P_{12} \leftarrow P_{14} P_{13}$)

Table 6.13 Effect of variation in parameter size (from 5, 10, 15, 20, and 22) with variable support (at confidence – 80%, 90%, and 100%) and corresponding number of association rules [**Category-B**]

Support (%)	1	2	3	4	5	10	20	30	40	50	60	70	80	90	100
ARM size (5/80%)	16	16	15	15	13	10	7	3	2	1	1	1	1	1	1
ARM size (10/80%)	2730	2730	2083	1185	827	81	13	5	2	1	1	1	1	1	1
ARM size (15/80%)	184,228	184,228	140,996	67,472	29,995	1479	201	60	20	16	12	11	4	4	3
ARM size (20/80%)	9,023,784	9,023,784	5,173,132	934,992	243,482	4326	349	101	44	36	22	11	4	4	3
ARM size (22/80%)	62,151,312	25,389,899	5,619,616	407,512	145,504	1995	118	39	15	11	7	3	1	1	0
ARM size (5/90%)	16	16	15	15	13	10	7	3	2	1	1	1	1	1	1
ARM size (10/90%)	2262	2262	1615	717	359	59	13	5	2	1	1	1	1	1	1
ARM size (15/90%)	8,864,766	8,864,766	5,014,114	775,884	84,464	2176	181	59	36	28	16	11	4	4	3
ARM size (20/90%)	8,864,766	8,864,766	5,014,114	775,884	84,464	2176	181	59	36	28	16	11	4	4	3
ARM size (22/90%)	62,050,814	25,289,401	5,519,118	307,014	45,006	734	35	14	9	7	4	3	1	1	0
ARM size (5/100%)	16	16	15	15	13	10	7	3	2	1	1	1	1	1	1
ARM size (10/100%)	2252	2252	1605	707	349	49	13	5	2	1	1	1	1	1	1
ARM size (15/100%)	166,558	166,558	123,326	49,802	12,325	439	65	22	10	8	6	5	2	2	1
ARM size (20/100%)	8,863,626	8,863,626	5,012,974	774,744	83,324	1036	97	33	18	14	8	5	2	2	1
ARM size (22/100%)	62,050,212	25,288,799	5,518,516	306,412	44,404	132	0	0	0	0	0	0	0	0	0

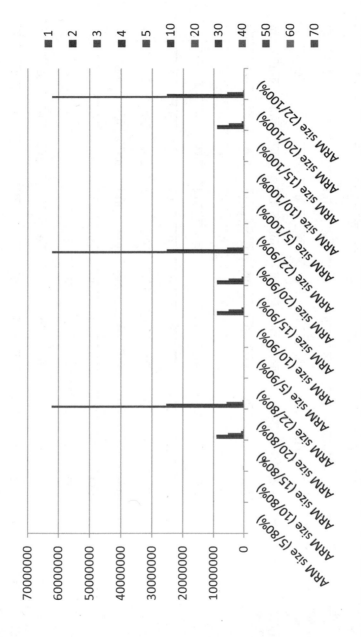

Figure 6.2 Effect of variation in parameter size (from 5, 10, 15, 20, and 22) with variable support (at confidence – 80%, 90%, and 100%) and corresponding number of association rules.

Table 6.14 Variation of support and confidence and time for generation of association rules (with 22 parameters) [Category C]

Support (%)	1	2	3	4	5	10	20	30	40	50	60	70	80	90	100
Time (5/80%)	0.12	0.02	0.05	0.05	0.05	0.01	0.05	0.05	0.05	0.05	0.05	0.05	0.05	0.05	0.05
Time (10/80%)	0.04	0.04	0.04	0.04	0	0	0	0	0	0	0	0	0	0	0
Time (15/80%)	0.17	0.19	0.13	0.06	0.03	0.05	0.04	0	0.06	0.04	0	0	0	0	0
Time (20/80%)	9.83	9.88	5.36	0.12	0.25	0.01	0	0	0	0	0	0	0	0	0
Time (22/80%)	108.27	27.21	5.67	0.46	0.15	0.02	0.01	0.01	0.01	0.01	0.01	0.01	0.01	0.01	0
Time (5/90%)	0.02	0.02	0.05	0.05	0.05	0.01	0.05	0.05	0.05	0.05	0.05	0.05	0.05	0.05	0.05
Time (10/90%)	0.04	0.04	0.04	0.04	0	0	0	0	0	0	0	0	0	0	0
Time (15/90%)	0.16	0.16	0.12	0.05	0.02	0.06	0.04	0	0.06	0.04	0	0	0	0	0
Time (20/90%)	9.61	9.51	5.26	0.77	0.08	0	0	0	0	0	0	0	0	0	0
Time (22/90%)	70.94	29.34	5.63	0.3	0.09	0.02	0.01	0.01	0.01	0.01	0.01	0.01	0.01	0.01	0
Time (5/100%)	0.02	0.02	0.05	0.05	0.05	0.04	0.05	0.05	0.05	0.05	0.05	0.05	0.05	0.05	0.05
Time (10/100%)	0.04	0.04	0.04	0.04	0	0	0	0	0	0	0	0	0	0	0
Time (15/100%)	0.16	0.16	0.19	0.05	0.01	0	0	0	0	0	0	0	0	0	0
Time (20/100%)	9.59	9.55	5.23	0.78	0.1	0	0	0	0	0	0	0	0	0	0
Time (22/100%)	71.91	28.41	5.62	0.3	0.06	0.01	0	0	0	0	0	0	0	0	0

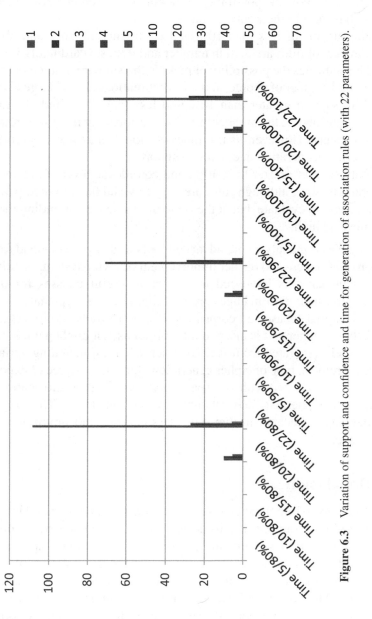

Figure 6.3 Variation of support and confidence and time for generation of association rules (with 22 parameters).

6.5.1 Observation for Association Rules Mining (ARM)

1. The rules show the possibility of picking the parameter combination and with respect to their correlation.
2. For support of less than 10% and confidence of 80% and above, the numbers of rules are high in number and item set cardinality is outsized. This indicates the possibility of picking the similar parameters for key by a small number of people, and the combinations are too large to evaluate.
3. With support greater than 10% and less than 40%, and confidence of 80% and above, the number of rules is rational but needs rigorous efforts to evaluate. However, its usefulness is doubtful as many parameters are out of consideration due to high support.
4. With support greater than 40%, and confidence level of greater than or equal to 80%, although rules are very few and have limited parameters, it is easy to evaluate, but at the same time, chances of finding the natural rules are high.

Enabling the cryptanalyst with advanced tools is always in demand for identification of weakness and further improvement of cryptosystem. In polynomial time, cryptanalyst is interested in extracting useful guesses for detecting original information, from huge corpus of ciphers. The cryptanalyst may have captured large database and corpus containing a variety of ciphers and hash files. When a cipher text is inserted into this dataset, it might get mixed within other ciphers generated from various other schemes, including variations in key size, protocol, type of cipher generation algorithm, degree of exposures of information concerned with key space, and lots of other information related to plaintext, ciphertext, and relationship between them. The cryptanalyst may develop a mechanism that will classify, sort, and group according to cipher type.

6.6 Conclusion

Success of any IoT application relies on the security of associated information and their testing frameworks. To design more secured and attack-proof IoT-enabled information system, more specific cryptographic approaches are needed. Also, data retrieval and processing are the integral parts of the whole IoT-based information system that need to be protected through high-level encryption. More advanced wireless protocols with strict keys are available for IoT security deployment. One such approach is based on the usage of an AVK, which is discussed in this chapter with a detailed analysis with variation in support and confidence. Although even after using the best encryption scheme, the device or network of the IoT system will remain vulnerable against the attacks. If there is no way to establish the authenticity of the data being communicated to and from an IoT device, the security may certainly be compromised. To resolve this issue, the AVK schemes may be useful for building. The proposed model has been tested and analyzed with multiple parameters, number of frequent sets, and number of association rules to be generated. The proposed AVK-based parametric model is user-friendly and yet harder from cryptic mining perspectives.

References

[1] Chakrabarti P., Bhuyan B., Chowdhuri A., and Bhunia C., A novel approach towards realizing optimum data transfer and Automatic Variable Key (AVK) in cryptography. *IJCSNS*. 8, 241, 2008.
[2] Chakrabarti, P., Bhuyan, B., Chowdhuri, A., and Bhunia, C. T., "Application of. Automatic Variable Key (AVK) in RSA," in *Int'l J HIT Transactions on ECCN*, 2, 304–311, 2007.
[3] Chakrabarti, P., Mondal, G. H., Bhunia, C. T., and Chowdhuri, A., Various New and Modified approaches for selective encryption (DES, RSA and AES) with AVK and their comparative study. *Int. J. Trans. ECCN*, 1(4), 236–244, 2008.
[4] Bhunia, C. T., Application of AVK and selective encryption in improving performance of quantum cryptography and networks, *United Nations Educational Scientific and Cultural Organization and International Atomic Energy Agency*, 10, 200–210, 2006.
[5] Dutta, M. P., Banerjee, S., and Bhunia, C., "Two New Schemes to Generate Automatic Variable Key (AVK) to achieve the Perfect Security in Insecure Communication Channel," in *Proceedings of the 2015 International Conference on Advanced Research in Computer Science Engineering and Technology (ICARCSET 2015)*, 1–4, 2015.
[6] Bhunia, C. T., Chakrabarti, P., Chowdhuri, A., and Chandan, T., Implementation of Automatic Variable Key with Choas Theory and Studied Thereof, *J. IUP Comp. Sci.* 5, 22–32, 2011.
[7] Bhunia, C. T., Mondal, G., and Samaddar, S., "Theories and Application of Time Variant Key in RSA and that with selective encryption in AES," in *Proc. EAIT, Elsevier Publications, Calcutta CSI-06*, 219–221, 2006.

[8] Bhunia, C. T., Chakrabarti, P., Goswami, R., "*A New Technique (CSAVK) of Automatic Variable Key in Achieving Perfect Security,*" 100th Indian Science Congress Association, 2013.

[9] Bhunia, C. T., New Approaches for Selective AES towards Trackling Error Propagation Effect of AES. *Asia. J. Inform. Technol.* 5, 1017–1022, 2006.

[10] Han, J., Kamber, M., and Pei, J., *Data Mining: Concepts and Technique*, 2nd edn, Morgan Kaufmann, 2006.

[11] Prajapat, S., and Thakur, R. S., Cryptic Mining: Apriori Analysis of Parameterized Automatic Variable Key based Symmetric Cryptosystem. *Int. J. Comp. Sci. Inform. Secur.* 14, 233–246, 2016.

[12] Prajapat, S., Rajput, D., and Thakur, R. S., "Time variant approach towards symmetric key," in *Proceedings of IEEE Science and Information Conference (SAI), London 2013*, 398–405, 2013.

[13] Prajapat, S., and Thakur, R. S.. "Optimal Key Size of the AVK for Symmetric Key Encryption," In *Covenant Journal of Information and Communication Technology*, 3(2), 71–81, 2015.

[14] Prajapat, S., Thakur, R. S., Various Approaches towards Crypt-analysis. *Int. J. Comp. Appl.* 127(14), 15–24, 2015.

[15] Prajapat, S., Thakur, R. S., Cryptic Mining for Automatic Variable Key Based Cryptosystem. *Elsevier Procedia Comp. Sci.* 78(78C), 199–209, 2016.

[16] Prajapat, S., Thakur, R. S., Realization of information exchange with Fibo-Q based Symmetric Cryptosystem. *Int. J. Comp. Sci. Inform. Secur.* 14(2), 216–223, 2016.

[17] Prajapat, S., Thakur, A., Maheshwari, K., and Thakur, R. S., Cryptic Mining in Light of Artificial Intelligence. *IJACSA* 6(8), 62–69, 2015.

Index

Glossary

Numbers

51% Attack
When more than half of the computing power of a cryptocurrency network is controlled by a single entity or group, this entity or group may issue conflicting transactions to harm the network, should they have the malicious intent to do so.

A

Address
Addresses (Cryptocurrency addresses) are used to receive and send transactions on the network. An address is a string of alphanumeric characters, but can also be represented as a scannable QR code.

Agreement Ledger
An agreement ledger is distributed ledger used by two or more parties to negotiate and reach agreement.

Attestation Ledger
A distributed ledger provides a durable record of agreements, commitments, or statements, providing evidence (attestation) that these agreements, commitments, or statements were made.

ASIC
ASIC is an acronym for "Application Specific Integrated Circuit." ASICs are silicon chips specifically designed to do a single task. In the case of bitcoin, they are designed to process SHA-256 hashing problems to mine new bitcoins.

B

Bitcoin (Uppercase)
It is the well-known cryptocurrency, based on the proof-of-work blockchain.

bitcoin (Lowercase)
It is the specific collection of technologies used by Bitcoin's ledger, a particular solution. Note that the currency is itself one of these technologies, as it provides the miners with the incentive to mine.

Blockchain
A blockchain is a type of distributed ledger, composed of unchangeable, digitally recorded data in packages called blocks (rather like collating them on to a single sheet of paper). Each block is then "chained" to the next block, using a cryptographic signature. This allows block chains to be used like a ledger, which can be shared and accessed by anyone with the appropriate permissions.

Block Height
Block height refers to the number of blocks connected together in the block chain. For example, Height 0, would be the very first block, which is also called the Genesis Block.

Block Reward
It is the reward given to a miner which has successfully hashed a transaction block. Block rewards can be a mixture of coins and transaction fees, depending on the policy used by the cryptocurrency in question, and whether all of the coins have already been successfully mined. The current block reward for the Bitcoin network is 25 bitcoins for each block.

C

Confirmation
It is the successful act of hashing a transaction and adding it to the blockchain.

Consensus
Consensus is achieved when all participants of the network agree on the validity of the transactions, ensuring that the ledgers are exact copies of each other.

Cryptocurrency
Also known as tokens, cryptocurrencies are representations of digital assets.
Cryptographic Hash Function
Cryptographic hashes produce a fixed-size and unique hash value from variable-size transaction input. The SHA-256 computational algorithm is an example of a cryptographic hash.

D

Dapp

A decentralized application (Dapp) is an application that is open source, operates autonomously, has its data stored on a blockchain, incentivized in the form of cryptographic tokens, and operates on a protocol that shows proof of value.

DAO

Decentralized autonomous organizations can be thought of as corporations that run without any human intervention and surrender all forms of control to an incorruptible set of business rules.

Distributed Ledger

Distributed ledgers are ledgers in which data are stored across a network of decentralized nodes. A distributed ledger does not have to have its own currency and may be permissioned and private.

Distributed Network

It is a type of network where processing power and data are spread over the nodes rather than having a centralized data center.

Difficulty

This refers to how easily a data block of transaction information can be mined successfully.

Digital Signature

It is a digital code generated by public key encryption that is attached to an electronically transmitted document to verify its contents and the sender's identity.

Double Spending

Double spending occurs when a sum of money is spent more than once.

E

Ethereum

Ethereum is a blockchain-based decentralized platform for apps that run smart contracts, and is aimed at solving issues associated with censorship, fraud, and third party interference.

EVM

The Ethereum Virtual Machine (EVM) is a Turing complete virtual machine that allows anyone to execute arbitrary EVM Byte Code. Every Ethereum node runs on the EVM to maintain consensus across the blockchain.

F

Fork
Forks create an alternate version of the blockchain, leaving two blockchains to run simultaneously on different parts of the network.

G

Genesis Block
It is the first or first few blocks of a blockchain.

H

Hard Fork
It is a type of fork that renders previously invalid transactions valid, and vice versa. This type of fork requires all nodes and users to upgrade to the latest version of the protocol software.

Halving

Bitcoins have a finite supply, which makes them a scarce digital commodity. The total amount of bitcoins that will ever be issued is 21 million. The number of bitcoins generated per block is decreased 50% every four years. This is called "halving." The final halving will take place in the year 2140.

Hash
It is the act of performing a hash function on the output data. This is used for confirming coin transactions.

Hash Rate
Measurement of performance for the mining rig is expressed in hashes per second.

Hybrid PoS/PoW
A hybrid PoS/PoW allows for both proof of stake and proof of work as consensus distribution algorithms on the network. In this method, a balance between miners and voters (holders) may be achieved, creating a system of community-based governance by both insiders (holders) and outsiders (miners).

I

Initial Coin Offering (ICO)
An initial coin offering (also called an ICO) is an event in which a new cryptocurrency sells advance tokens from its overall coinbase, in exchange for upfront capital. ICOs are frequently used for developers of a new cryptocurrency to raise capital.

L

Ledger
It is an append-only record store, where records are immutable and may hold more general information than financial records.

Litecoin
It is a peer-to-peer cryptocurrency based on the Scrypt proof-of-work network. Sometimes it is referred to as the silver of bitcoin's gold.

M

Mining
Mining is the act of validating blockchain transactions. The necessity of validation warrants an incentive for the miners, usually in the form of coins. In this cryptocurrency boom, mining can be a lucrative business when done properly. By choosing the most efficient and suitable hardware and mining target, mining can produce a stable form of passive income.

Multi-Signature
Multi-signature addresses provide an added layer of security by requiring more than one key to authorize a transaction.

N

Node
It is a copy of the ledger operated by a participant of the blockchain network.

O

Off-Ledger Currency
It is a currency minted off-ledger and used on-ledger. An example of this would be using distributed ledgers to manage a national currency.

On-Ledger Currency
It is a currency minted on-ledger and used on-ledger. An example of this would be the cryptocurrency, Bitcoin.

Oracles
Oracles work as a bridge between the real world and the blockchain by providing data to the smart contracts.

P

Peer to Peer
Peer to peer (P2P) refers to the decentralized interactions between two parties or more in a highly interconnected network. Participants of a P2P network deal directly with each other through a single mediation point.

Public Address
A public address is the cryptographic hash of a public key. They act as email addresses that can be published anywhere, unlike private keys.

Private Key
A private key is a string of data that allow you to access the tokens in a specific wallet. They act as passwords that are kept hidden from anyone but the owner of the address.

Proof of Stake
It is a consensus distribution algorithm that rewards earnings based on the number of coins you own or hold. The more you invest in the coin, the more you gain by mining with this protocol.

Proof of Work
It is a consensus distribution algorithm that requires an active role in mining data blocks, often consuming resources, such as electricity. The more "work" you do or the more computational power you provide, the more coins you are rewarded with.

R

Ripple
It is a payment network built on distributed ledgers that can be used to transfer any currency. The network consists of payment nodes and gateways operated by authorities. Payments are made using a series of IOUs, and the network is based on trust relationships.

Replicated Ledger
It is a ledger with one master (authoritative) copy of the data and many slave (non-authoritative) copies.

S

Scrypt
Scrypt is a type of cryptographic algorithm and is used by Litecoin. Compared to SHA256, this is quicker as it does not use up as much processing time.

SHA-256
SHA-256 is a cryptographic algorithm used by cryptocurrencies such as Bitcoin. However, it uses a lot of computing power and processing time, forcing miners to form mining pools to capture gains.

Smart Contracts
Smart contracts encode business rules in a programmable language onto the blockchain and are enforced by the participants of the network.

Soft Fork
A soft fork differs from a hard fork in that only previously valid transactions are made invalid. Since old nodes recognize the new blocks as valid, a soft fork is essentially backward-compatible. This type of fork requires most miners upgrading in order to enforce, while a hard fork requires all nodes to agree on the new version.

Solidity
Solidity is Ethereum's programming language for developing smart contracts.

T

Testnet
It is a test blockchain used by developers to prevent expending assets on the main chain.

Transaction Block
It is a collection of transactions gathered into a block that can then be hashed and added to the blockchain.

Transaction Fee
All cryptocurrency transactions involve a small transaction fee. These transaction fees add up to account for the block reward that a miner receives when he successfully processes a block.

Turing Complete
Turing complete refers to the ability of a machine to perform calculations that any other programmable computer is capable of. An example of this is the Ethereum Virtual Machine (EVM).

U

Unpermissioned ledgers
Unpermissioned ledgers such as Bitcoin have no single owner – indeed, they cannot be owned. The purpose of an unpermissioned ledger is to allow anyone to contribute data to the ledger and for everyone in possession of the ledger to have identical copies. This creates censorship resistance, which means that no actor can prevent a transaction from being added to the ledger. Participants maintain the integrity of the ledger by reaching a consensus about its state.

W

Wallet

It is a file that houses private keys. It usually contains a software client which allows access to view and create transactions on a specific blockchain that the wallet is designed for.

References

[1] Available at: https://blockgeeks.com/guides/blockchain-glossary-from-a-z/
[2] Available at: http://www.blockchaintechnologies.com/blockchain-glossary

About the Editors

Dr. Shishir Kumar Shandilya is a renowned academician and an active researcher with proven record of teaching and research. He is a Cambridge University Certified Professional Teacher and Trainer, a Senior Member of IEEE-USA, and also elected as an executive member of the IEEE Industry-Outreach Committee-India. Dr. Shandilya has received the "IDA Teaching Excellence Award" for distinctive use of Technology in Teaching by Indian Didactics Association, Bangalore, and the "Young Scientist Award" for two consecutive years (2005 and 2006) by Indian Science Congress and MP Council of Science and Technology. He has written seven books of international-fame (published in USA, Denmark, and India) and published over 50 quality research papers. He is an active member of over 20 international professional bodies. He is also an excellent programmer and credited various software projects in his account. He is also giving consultancy in IT as a Sr. Consultant. He has recently delivered an expert lecture on "Opinion Mining" at Oxford – United Kingdom.

Dr. Soon Ae Chun is a professor and the director of Information Systems and Informatics program, and a Graduate Center doctoral faculty member of Computer Science at the City University of New York. She is the director of the Information Security Research and Education Lab (iSecure Lab) sponsored by NSF. She is a Fulbright Scholar and the recipient of the CSI Dolphin Award for Outstanding Scholarly Achievement. She is served as the President of the Digital Government Society. Professor Chun's research interests include security and privacy, semantic web, data integration, social data analytics, and workflow. Dr. Chun's current research projects include the development of a cybersecurity ontology, a Linked Data development of multi-modal cybersecurity educational data, and research on social data analytics in the Healthcare domain. Her research has been sponsored by NSF, NOAA, a New Jersey State Agency, CUNY, and Fulbright. Professor Chun has published in security and database journals, such as Journal of Computer Security, IEEE Transactions on Dependable and Secure Computing, Journal of Distributed and Parallel Databases, Journal of Information Sciences, as well as applied informatics journals such as Government Information Quarterly and Information Polity. Dr. Chun has served as conference chairs at the International Conference on Digital Government Research and the IFIP conference of Database and Applications Security and Privacy, a track chair of the Semantic Web Applications at ACM Symposium of Applied Computing, and a program chair of the International Conference on Digital Government Research. Professor Chun is an IEEE senior member and a member of the ACM, the Digital Government Society, and the Beta Gamma Sigma Honor Society.

Dr. Smita Shandilya is an eminent scholar and energetic researcher with excellent teaching and research skills. She achieved excellent result in all the subjects she has taught till date. She has over 20 quality research papers in international and national journals and conferences to her credits. She has delivered several invited talks in national seminars of high repute. Her research interests are power system planning and smart micro grids. She is one of the core members of the research and development section of her Institute. She is also involved in various projects like the establishment of Energy Lab in the Institute (first in any Private Institute in Madhya Pradesh), Establishment of Training cum Incubator center in Collaboration with iEnergy.

Prof. Edgar Weippl is the research director of SBA Research and an Associate Professor (Privatdozent) at the TU Wien. His research focuses on applied concepts of IT-security and e-learning. After graduating with a Ph.D. from the TU Wien, Edgar worked in a research startup for two years. He then spent one year teaching as an Assistant Professor at Beloit College, WI. From 2002 to 2004, while with the software vendor ISIS Papyrus, he worked as a consultant in New York, NY, and Albany, NY, and Frankfurt, Germany. In 2004, he joined the TU Wien and founded the research center SBA Research together with A Min Tjoa and Markus Klemen. Edgar R. Weippl (CISSP, CISA, CISM, CRISC, CSSLP, CMC) is a member of the editorial board of Computers and Security (COSE), organizes the ARES conference, and is the General Chair of SACMAT 2015, the PC Chair of Esorics 2015, the General Chair of ACM CCS 2016, and the PC Chair of ACM SACMAT 2017.